Bioinformatics Toolkit Decoding Proteomics

ADROIT
ADROIT PUBLISHERS
New Delhi, INDIA

Bioinformatics Toolkit Decoding Proteomics

Title : Bioinformatics Toolkit Decoding Proteomics

Author : Dr Uma Kumari

Year : 2025 (1st Edition)

ISBN : 978-93-48292-44-5

Published by :
ADROIT PUBLISHERS 4675/21, Ganpati Bhawan, Ansari Road,
Daryaganj, New Delhi-110002, INDIA, Phone : 011-45583612
adroitpublishers@gmail.com

Typeset by Gaurav Sati,
Ph. 9289951990
E-mail : gsati010@gmail.com

Printed by: Impackt Printing & Lamination
Mob. 7827944816

Contents

Preface

With great enthusiasm, I present Bioinformatics **Toolkit: Decoding Proteomics (A Practical Approach)** a book shaped by my passion, experience, and commitment to making bioinformatics accessible, especially in proteomics. The rapid decoding of complete genome sequences of humans, plants, and microorganisms has opened vast research avenues through comparative and functional proteomics. My work from 2013 to 2025 on cancer proteomics, genomics, neurodegenerative research, NGS data analysis, and drug discovery exposed both the complexity and the transformative potential of computational biology. This journey, strengthened by extensive reading and hands-on research, inspired me to simplify these concepts for learners. My teaching experience from MSc and BSc programs at Ranchi University to B.Tech, M.Tech, PhD, postdoctoral scholars, and professionals at the Bioinformatics Project and Research Institute, Noida revealed a clear gap: while information on bioinformatics is abundant, practical, step-wise, proteomics focused guidance is scarce.This book bridges that gap. Each chapter provides clear explanations, protocols, case studies, and step-by-step procedures to help students and researchers confidently apply bioinformatics tools in real-world proteomic analysis. It is designed as a practical companion for anyone entering or advancing in computational biology. I express heartfelt gratitude to my parents, family, friends, students, and the contributing authors for their constant support and inspiration.

Dr. Uma Kumari

Foreword

Research in Bioinformatics is undergoing tremendous changes observe in year 2019 to 2025.In the ever-evolving landscape of life sciences, proteomics has emerged as a pivotal field unveiling the intricate architecture of life through the study of proteins, their structures, functions, and interactions. As biological data continues to expand exponentially, the role of bioinformatics in decoding this complexity has never been more critical.

Bioinformatics Toolkit: Decoding Proteomics stands as a timely and essential guide for students, researchers, and professionals navigating this complex yet fascinating domain. This book provides more than just a collection of tools it offers a roadmap to understanding how computational resources and proteomic data can be effectively applied to real-world proteomics challenges.

What makes this work truly valuable is its clarity and practicality. Each chapter is designed to bridge the gap between theory and application, enabling readers to not only understand the tools but also appreciate their significance in biological discovery. From protein identification and structural prediction to interaction mapping and pathway analysis, this toolkit empowers readers to uncover the hidden stories encoded within proteins.

As the scientific community moves toward integrative and personalized approaches to health, agriculture, and biotechnology, mastering the synergy between proteomics and bioinformatics is no longer optional it's essential.Whether you are a budding bioinformatician, a seasoned proteomics researcher, or someone curious about the digital language of life, this book will serve as both a foundation and an inspiration. Let this toolkit be your guide as you decode the proteome and in doing so, decode life itself.(In terms of next generation sequence analysis ,mutational analysis,and many more)

Dr Uma Kumari

Founder, Director, Bioinformatics Scientist

BIOINFORMATICS PROJECT AND RESEARCH INSTITUTE, NOIDA,UP 201301

Acknowledgment

The completion of this book, Bioinformatics Toolkit: Decoding Proteomics, has been made possible through the collective support, guidance, and encouragement of many individuals and institutions who contributed directly and indirectly to its realization.I express my sincere gratitude to the scientific community of bioinformatics and proteomics researchers whose pioneering work from 2019 to 2025 has laid the foundation for the concepts, tools, and methodologies discussed in this book. Their continuous efforts in advancing computational biology have been a constant source of inspiration.

I would like to acknowledge the unwavering support of the **Bioinformatics Project and Research Institute (BPRI), Noida**, whose academic environment, research infrastructure, and commitment to interdisciplinary learning enabled the successful development of this work. The institute's emphasis on integrating theory with practical applications greatly influenced the structure and intent of this book.

My heartfelt thanks go to students, research scholars, and professionals who through their curiosity and real-world challenges, motivated the creation of a resource that bridges the gap between proteomics data and computational interpretation. Their questions and feedback helped shape this toolkit into a practical guide rather than a purely theoretical text.

I extend my sincere gratitude to my father Dr Rajendra Prasad Sahu Mother Lal Mani Devi, Spouse Pravin Kumar,my daughter Aavya,Elder Brother Awinash kumar Gupta, Anant Kumar, my best friend Meenakshi Pradhan and my all students who have been part of this journey. Their insights and encouragement have greatly enriched this work.

Dr. Uma Kumari
Founder & Director | Bioinformatics Scientist
Bioinformatics Project and Research Institute (BPRI)
Noida, Uttar Pradesh, India

Introduction

Bioinformatics Toolkit: Decoding Proteomics

Biological research has undergone a remarkable transformation over the last few decade. The advancement of sequencing platforms, imaging tools, and spectrometric techniques has resulted in an unprecedented generation of biological data. While this abundance of data has opened new doors to discovery, it has also introduced significant challenges in data storage, organization, and interpretation. Addressing these challenges has led to the emergence of bioinformatics as an essential scientific field. Bioinformatics integrates biology with computational science to analyse molecular data and convert raw information into biological knowledge. It enables researchers to examine genetic sequences, predict protein structures, identify disease markers, and analyze complex metabolic pathways. Without computational tools, managing modern biological datasets would be nearly impossible. Databases, statistical models, and algorithms now play an equally important role as microscopes and experimental reagents in biological laboratories. Among various disciplines within bioinformatics, proteomics occupies a critical position. While genetic information provides a blueprint of life, proteins are the functional units that drive biological processes. Enzymatic reactions, cellular signalling, immune responses, and metabolic activities are executed by proteins. Unlike the genome, which remains relatively constant, the proteome is dynamic and changes with environmental conditions, developmental stages, and disease states. Understanding proteins at the molecular level is therefore essential for accurate biological interpretation.

With the continuous development of computational tools and experimental methods, proteomic research has expanded rapidly. High-throughput technologies generate vast datasets that require efficient computational analysis. Software applications, structural modelling platforms, and machine learning-based systems assist researchers in translating experimental observations into meaningful conclusions. Bioinformatics has thus become the link connecting molecular biology with real-world applications in medicine, agriculture, and biotechnology. This book is designed to present bioinformatics as a practical and analytical discipline rather than a purely theoretical subject. It aims to build conceptual knowledge while encouraging hands-on learning. Each chapter provides clarity on fundamental principles, supported by practical examples to enhance understanding. Special attention has been paid to presenting complex topics in a simplified manner without compromising scientific depth.

The motivation behind writing this book lies in the growing demand for trained professionals in bioinformatics and computational biology. Although numerous learning resources are available, learners often encounter difficulties due to fragmented content and inconsistent teaching approaches. This book addresses that gap by offering a structured progression from introductory topics to advanced techniques.

The early chapters introduce fundamental concepts such as sequence formats, similarity analysis, and alignment methods. These topics serve as the foundation for studying structural analysis, homology modelling, and proteomic workflows. Readers are guided through essential biological databases and digital resources that are widely used in academic and industrial research. Practical tools are discussed alongside explanations of their real-world significance.

As research increasingly relies on computational modelling and predictive analytics, the book includes a dedicated focus on artificial intelligence and machine learning as applied to biological problems. Algorithms that were once limited to engineering disciplines are now powerful tools in biological data interpretation. These computational approaches have revolutionized areas such as drug discovery, disease detection, and genome annotation.

Next-generation sequencing has been presented as a cornerstone of modern genomic research. This book explores the principles underlying sequencing technologies and examines how sequencing data is processed, analysed, and interpreted. The applications of sequencing extend beyond medical research into fields such as environmental science, food biotechnology, and microbial ecology.

Food biotechnology represents another important domain addressed in this book. The application of bioinformatics in food research has enabled improved nutritional quality, assessment of food safety, and development of functional foods. Techniques such as metagenomics have expanded our understanding of microbial communities responsible for fermentation and probiotic development.

Structural bioinformatics has also been emphasized due to its significance in understanding molecular interactions. Protein structures reveal important information regarding function, stability, and binding capability. Computational modelling allows the visualization of complex molecular systems that are otherwise invisible at the experimental level. A distinct feature of this book is the inclusion of case studies that reflect real-world biological applications. These examples allow readers to understand how theoretical knowledge is implemented in research and industry settings. Topics such as COVID-19 are used to demonstrate how bioinformatics contributes to global health challenges.

Career opportunities in bioinformatics continue to expand as industries integrate computational biology into their research workflows. This book provides information on professional roles, required skills, and learning pathways for students interested in pursuing careers in this field. The intention of this work is not merely to describe bioinformatics tools but to foster analytical thinking and scientific curiosity. Readers are encouraged to explore biological questions using computational approaches and develop solutions through interdisciplinary learning.

In conclusion, this book seeks to bridge the gap between experimental biology and computational analysis. It aims to empower learners with knowledge, practical understanding, and confidence to engage with modern biological research. As life sciences continue to evolve, bioinformatics will remain a cornerstone of scientific advancement, and it is hoped that this book will inspire readers to become contributors to that progress.

1 Introduction of Bioinformatics

Abstract

Bioinformatics has emerged as a pivotal field for managing, analyzing, and interpreting biological data. Its success largely depends on a wealth of resources, including databases, software tools, and computational platforms. This chapter explores the major bioinformatics resources, highlighting their roles in advancing biological research. It discusses sequence, structure, and functional annotation databases as well as computational tools and web-based platforms that facilitate genome analysis, proteomics, transcriptomics, and systems biology.

Introduction

To analyze and interpret biological data, the field of bioinformatics integrates biology, computer science, mathematics, and statistics. It entails the creation and use of computational methods and tools to comprehend biological processes, forecast the structures of proteins, examine DNA sequences.

Bioinformatics plays a crucial role in advancing our understanding of various biological phenomena by harnessing computational methods to analyze vast amounts of biological data. From deciphering the human genome to studying the interactions of proteins within cells, bioinformatics provides valuable insights into complex biological systems. By using computer techniques to analyses enormous volumes of biological data, bioinformatics significantly advances our understanding of a variety of biological phenomena. It offers important insights into intricate biological processes, ranging from understanding the human genome to studying the relationships between proteins in cells (Durbin *et al.,1998)*

What is Bioinformatics?

Bioinformatics, which Combining biology and information technology, involves the collection, classification, storage, and computer-based analysis of biological and biochemical data, particularly as it relates to molecular genetics and genomics.

Bioengineering and biology are combined in biological computation to create computational analysis. On the other hand, bioinformatics makes use of computational technologies to comprehend biology better. Computation and bioinformatics Analysis of biological data, especially DNA, RNA, and protein sequences is a core component of bioinformatics

Bioinformatics is the application of computer technology to get the information that's stored in certain types of biological data server.

Bioinformatics provides central, globally accessible databases that enable scientists to submit, search and analyze information. It offers analysis software for data studies and comparisons and provides tools for modelling, visualizing, exploring and interpreting data. Main goal is to convert multitude of complex data into useful information and knowledge.

Scope in Bioinformatics

The Scope of Bioinformatics in different specialized areas like Current biological and medical labs use data analysis methods that produce extremely large data sets, which cannot be analyzed by hand - for instance sequencing human genomes. Thus, modern biological and medical research and development cannot be done without bioinformatics

Bioinformatics is an interdisciplinary field that integrate biology, computing, and informatics to analyze and interpret biological data. The scope of bioinformatics is broad and will continue to expand as technology advances and the amount of biological information increases.

Core Areas Where Bioinformatics Play A Critical Role

Genomics and Transcriptomics: Bioinformatics is important for the analysis of genomic and transcriptomic data, including genome sequencing, gene expression profiling, and comparative genomics.

Proteomics: analysis of protein sequences and structures, study of protein interactions and identification of potential drug targets.

Medical and Clinical Applications: Development of personalized medical approaches, analysis of patients' genetic data to predict disease risk, and identification of biomarkers for diagnosis and treatment.

Evolutionary Biology: Study of evolutionary relationships, creation of phylogenetic trees and understanding of the genetic basis of evolution and species differences.

Agricultural Biotechnology: Improvement of crops, development of disease resistant plants and research of plant genomes for better agricultural practices.

Pharmaceutical R&D: Identification of new drug targets, prediction of drug interactions and virtual screening of drug candidates.

Systems Biology: Integrating different types of biological data to model and understand complex biological systems and networks.

Structural Biology: The analysis of the three-dimensional structures of biomolecules and the understanding of their functions and interactions.

Data Management and Integration: Development of databases and tools to store, retrieve and integrate large-scale biological data.

Computational Biology: Applying algorithms, statistical models and machine learning techniques to solve biological problems. Bioinformatics professionals work in academia, industry, healthcare and government, advancing research, medicine and biotechnology. As biological data continues to explode, the demand for bioinformatics expertise and innovative solutions remains high, making this a dynamic and influential field.

Bioinformatics Applications in Biology, chemistry, pharmaceuticals, medicine, and agriculture.

In addition, bioinformatics plays an important role in biomedical research. Research work in the area of genetic diseases and medical genomics is rapidly increasing and the future of personalized medicine depends on bioinformatics approaches.

Bioinformatics is the application of computer technology to get the information that's stored in certain types of biological data. Bioinformatics provides central, globally accessible databases that enable scientists to submit, search and analyze information.

History of Bioinformatics

foundations of bioinformatics were laid in the early ١٩٦٠s with the application of computational methods to protein sequence analysis. **Paulien Hogeweg** and **Ben** Hesper coined it in ١٩٧٠ to refer to the study of information processes in biotic systems.

David Lipman, director of the National Center for Biotechnology Information, has called **Dayhoff** the **mother** and father of bioinformatics.

Dayhoff gathered all the available sequence **data** to create the **first bioinformatic database**. The Protein **Data** Bank followed in ١٩٧٢ with a collection of ten X-ray crystallographic protein structures.

Career Options in Bioinformatics

Bioinformatics is a rapidly growing career field and an emerging scientific discipline. This course focuses on employing existing bioinformatic resources - mainly web-based programs and databases - to access the wealth of data to answer questions relevant to the average biologist, and is highly hands-on.

There are different types of career opportunities available for different stream students, Scientific Curator, Gene Analyst, Protein Analyst, Phylogenetic, Research Scientist / Associate, Data base programmer, Bioinformatics software developer, Computational biologist, Network Administrator / Analyst, Structural analyst, Database programmer, Cheminformatician, Pharmacogenetician, Pharmacogenomics, Research Scientist / Associate.

What are the Job roles?

- Professor
- Bioinformatics Trainer/ Research Associate
- Computational Biologist
- Science Technician
- Research Assistant
- Bioinformatics Scientist
- Bioinformatics Analyst
- Junior Research Fellow
- Research Associate
- Bioinformatics Software Developer
- and foreign universities.

Major Resources and Database of Bioinformatics

- **NCBI -** National Center for Biotechnology Information
- **EBI -** European Bioinformatics Institute
- **DDBJ -** DNA Data Bank of Japan

Protein Sequence Databases

- **SWISS-PROT & TREMBL** - Protein sequence database and computer annotated supplement
- **UniProt** - UniProt (Universal Protein Resource) is the world's most comprehensive catalog of information on proteins. It is a central repository of protein sequence and function created by joining the information contained in Swiss-Prot, TrEMBL, and PIR.
- **PIR** - Protein Information Resource
- **MIPS** - Munich Information Centre for Protein Sequences

Database Searching by Sequence Similarity

- BLAST @ NCBI

NCBI (National center of Biotechnology Information)

TheNational Center for Biotechnology Information(NCBI) (Altschul Stephen et al., 1990) is part of theUnited States National Library of Medicine (NLM), a branch of theNational Institutes of Health(NIH). The NCBI is located in Bethesda, Maryland and was founded in 1988 through legislation sponsored by Senator Claude Pepper.

The NCBI houses a series of databases relevant to biotechnology and biomedicine and is an important resource for bioinformatics tools and services. Major databases include GenBank for DNA sequences and PubMed, a bibliographic database for biomedical literature. Other databases include the NCBI Epigenomics database. All these databases are available online through the Entrez search engine. NCBI was directed by David Lipman, one of the original authors of the BLAST sequence alignment program and a widely respected figure in bioinformatics. He also led an intramural research program, including groups led by Stephen Altschul (another BLAST co-author), David Landsman, Eugene Koonin, John Wilbur, Teresa Przytycka, and Zhiyong Lu. David Lipman stood down from his post in May 2017. NCBI for all the major databases such as Nucleotide and Protein Sequences, Protein Structures, PubMed, Taxonomy, Complete Genomes, OMIM, and several others in 2012

Gene has been implemented at NCBI to characterize and organize the information about genes. It serves as a major node in the nexus of the genomic map, expression, sequence, protein function, structure, and homology data. Gene records for known or predicted genes are established here and are demarcated by map positions or nucleotide sequences. Gene has several advantages over its predecessor, Locus Link, including, better integration with other databases.

Application of Bioinformatics in Biotechnology

Bioinformatics plays a vital role in the management of large volumes of genomic and/or proteomic data worldwide. It is used to design drugs, to model homology, to predict motifs/ domains, to document, to simulate dynamic simulations, and to manage DNA/proteomic sequences. Bioinformatics has a wide range of applications in various fields, such as: Genome sequencing Proteomics Nutrigenomics Comparative Genomics Microbial genomes Biodefense

Forensics Biosciences have a promising future in advancing biotechnological research. DNA sequences need to be preserved and analyzed from various organisms in order to be able to analyze them. In recent years, advances in structural/functional genomics and molecular technologies such as genome sequencing, DNA microarray have provided valuable insights into nutrition related to an individual's genetics, which has a direct impact on their metabolism. As a result, there has been a dramatic increase in the number of nutrition-related

Worldwide, the vast amounts of genomic and proteomic data are managed by bioinformatic tools and software, which include drug design, homology modelling, motif/domain prediction, documenting, dynamic simulation, and managing DNA/protein sequences. There are many uses for bioinformatics across many fields, including genomics. Analysis of proteomics, nutrigenomics, comparative genomics, microbial genomes, biodefense, forensics, etc. As a result, it has a bright future for advancing biotechnology research. To preserve and analyze DNA sequences from various organisms, genomic data and bioinformatic tools are needed (**J Y Yang et al., 2008**).

Protein structure predictions are now much easier to analyze, so bioinformatics can be used to predict 3D protein biological sample structure (LA Kelley et al.,2009) Advances in structural / functional genomics, and molecular technologies like genome sequencing, and DNA microarray, provide valuable insights that explain nutrition in relation to an individual's genetics, which directly affects their metabolism. Due to the increasing availability of bioinformatic tools, there is a huge increase in nutrition-related research Corthész-Theulas et al., 2005

Major Resources of Bioinformatics

Nucleotide Sequence Databases

GenBank: A comprehensive database maintained by NCBI, containing annotated nucleotide sequences from various species.

European Nucleotide Archive (ENA): Managed by EMBL-EBI, providing access to raw sequence data and functional annotations.

DDBJ: A Japanese counterpart of GenBank, forming part of the International Nucleotide Sequence Database Collaboration.

Protein Sequence Databases

UniProt: A curated repository of protein sequences and annotations, including UniProtKB, UniRef, and UniParc.

PIR (Protein Information Resource): Offers protein sequence and function information, particularly useful for functional genomics studies.

Structural Databases

Structural databases provide three-dimensional information about biological macromolecules, aiding drug design and protein function analysis.

Protein Data Bank (PDB): A repository of 3D structures of proteins, nucleic acids, and complex assemblies.

For example, home page of PDB

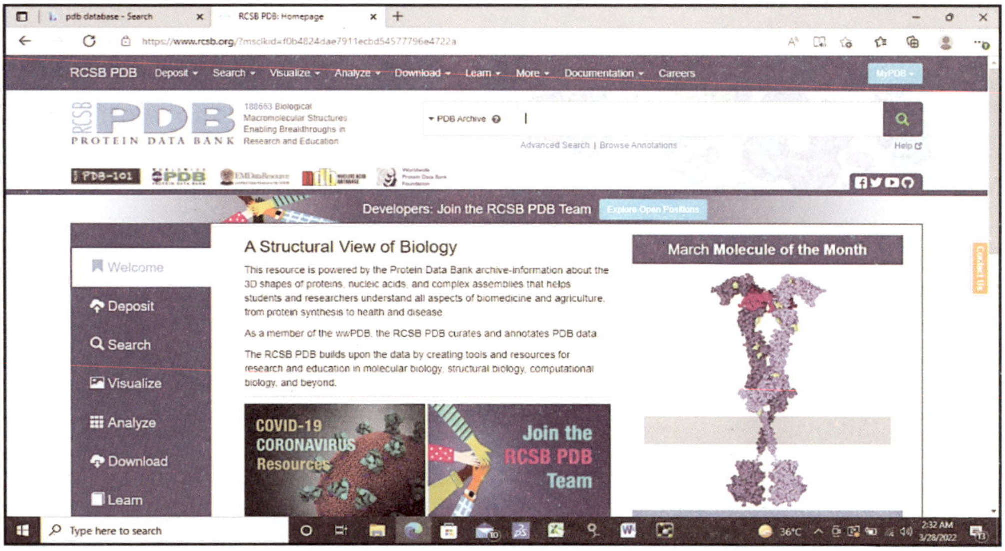

Figure 4: Homepage of Protein Data Bank

MMDB: Molecular Modelling Database at NCBI integrates structural data with biological annotations. In Entrez's the Structure' database correspond to the is the Molecular Modeling Database (MMDB). Experimental 3D structural data are retrieved from the protein data bank. For each structure, the agreement of atomic coordinate and sequence data is examined and sequences are automatically adjusted when necessary to ensure precise agreement with the coordinates. MMDB presently comprises around 20000 structural entries, correspond to about 40000 chains and 70000 3D domains. Sequences obtained from MMDB are placed into Entrez's protein or nucleic acid sequence databases, with a connection to the appropriate structure maintained. In MMDB, each 3D domain has approximately 600 structural neighbours on average (Wang Y *et al.,* 2002).

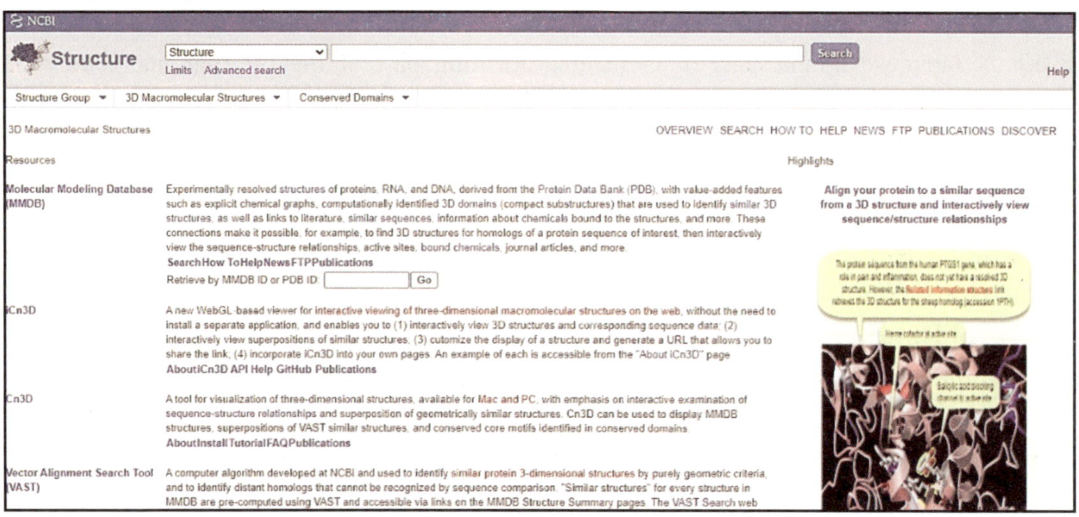

Figure 8: Homepage of Molecular Modeling Database

Functional Annotation Resources

Functional annotation tools and databases facilitate the understanding of gene and protein functions.

Gene Ontology (GO): A hierarchical system for annotating genes and proteins with functions, biological processes, and cellular components.

KEGG (Kyoto Encyclopaedia of Genes and Genomes): Provides pathway maps and annotations for metabolic, signalling, and disease processes.

Specialized Databases

Genomics

ENSEMBL: A genome browser providing integrated data from genome projects.

NCBI Genome: Offers complete genome sequences and annotations.

Proteomics

PRIDE: A proteomics database for storing mass spectrometry-based data.

Peptide Atlas: Contains high-quality proteomics data mapped to reference sequences.

Transcriptomics

Array Express: Stores functional genomics data from microarray and RNA-seq experiments.

GEO (Gene Expression Omnibus): A public repository for gene expression data.

Computational Tools and Web Based Platform

Bioinformatics tools and platforms provide computational capabilities for analyzing and visualizing biological data.

BLAST (Basic Local Alignment Search Tool): Facilitates sequence alignment and similarity searches.

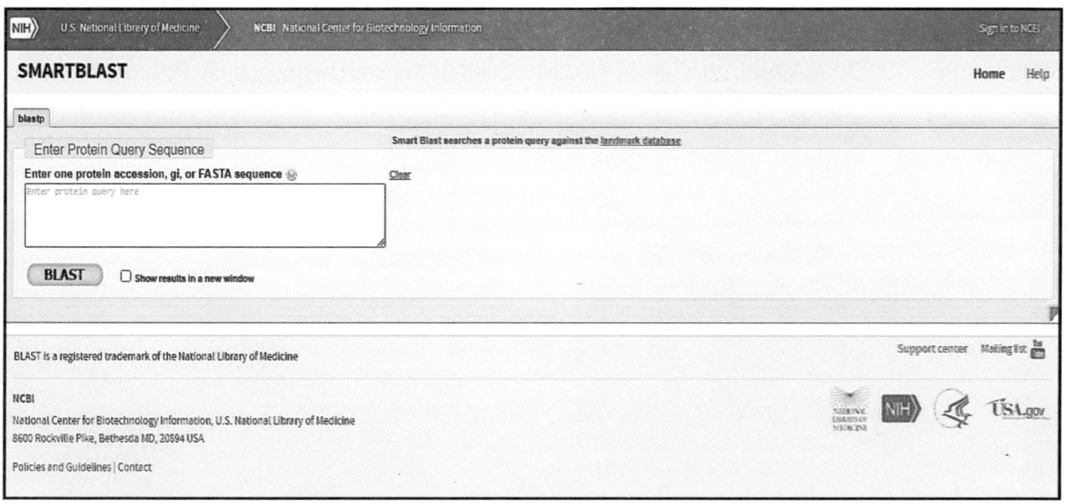

Figure: Blast home page for sequence similarity search

Crustal Omega: A tool for multiple sequence alignment.

Challenges and Future Direction

While bioinformatics resources have revolutionized biological research, challenges such as data integration, standardization, and scalability persist. Future directions include developing AI-driven tools, enhancing data interoperability, and creating more user-friendly platforms for non-expert users.

Conclusion

The rich array of bioinformatics resources has transformed our ability to analyze biological systems, driving discoveries in genomics, proteomics, and systems biology. Continuous advancements in these resources are essential for addressing emerging challenges in life sciences. To sum up, the multidisciplinary area of bioinformatics combines information technology, computer science, statistics, and biology to analyse and understand biological data. In Analysis of Data Large biological datasets, such as gene expression profiles, protein structures, genomic sequences, and more, are analyzed in bioinformatics. Algorithms and Tools: Bioinformaticians create and apply algorithms and computer tools for processing, analyzing, and interpreting biological data. These tools include machine learning models, statistical procedures, protein structure prediction approaches, and sequence alignment, algorithms. Bioinformatics places a lot of emphasis on genomics, which includes variant analysis, genome assembly, annotation, comparative genomics, and the study of DNA sequences. The study of proteins also makes use of bioinformatics, which is used in protein sequence analysis, structure prediction,

Refrence

1. Durbin, R., Eddy, S. R., Krogh, A., & Mitchison, G. (1998). *Biological Sequence Analysis: Probabilistic Models of Proteins and Nucleic Acids.* Cambridge University Press.(https://www.cambridge.org/core/books/biological-sequence analysis/50B888FDBA01B3E805A60A6F39C4973A)
2. Mount, D. W. (2004). *Bioinformatics: Sequence and Genome Analysis. Cold Spring Harbor Laboratory Press. (https://www.cshlpress.com/default.tpl?action=full&cart=16104509686817837 4&--eqskudatarq=956)
3. Baxevanis, A. D., & Ouellette, B. F. (Eds.). (2005). *Bioinformatics: A Practical Guide to the Analysis of Genes and Proteins.* John Wiley & Sons. (https://www.wiley.com/en-us/

2 Sequence Format

Introduction

What is a Sequence Format?

In Sequence format allowed structure and content of text in a file are defined by the Bioinformatics Study of Sequence Format. Text tokens that describe fields utilized in a databank are included in this. These fields contain, among other things, the accession number, the sequence identifier name, and the sequence itself. Most formats may be viewed on screen or printed out because non-printable control characters are not typically utilized.

One format that is frequently used format is FASTA. It is composed of a header line that begins with the ">" character, a code that identifies the sequence and frequently some text that describes the sequence. One or more lines comprising the sequence itself come after the header line. Sequences can be found in FASTA files in several copies.

Plain Sequence Format

A sequence in plain format may contain only IUPAC characters and spaces (no numbers!).

Note: The Plain sequence format may only contain **one** sequence, whereas most other formats can include multiple sequence within a single file.

An example sequence in plain format is:

A C A A G A T G C C A T T G T C C C C C G G C C T C C T G C T G C T G C T
G C T C T C C G G G G C C A C G G C C A C C G C T G C C C T G C C
C C T G G A G G G T G G C C C C A C C G G C C G A G A C A G C G A G C A T A T G C
A G G A A G C G G C A G G A A T A A G G A A A A G C A G C

C T C C T G A C T T T C C T C G C T T G G T G G T T T G A G T G G A C C T C C C A G
G C C A G T G C C G G G C C C C T C A T A G G A G A G G

A A G C T C G G G A G G T G G C C A G G C G G C A G G A A G G C G C A C C C C C C
C A G C A A T C C G C G C G C C G G G A C A G A A T G C C

C T G C A G G A A C T TC T T C T G G A A G A C C T T C T C C T C C T G C A A A T A A
A A C C T C A C C C A T G A A T G C T C A C G C A A G

T T T A A T T A C A G A C C T G A A

FASTQ Format

A sequence file in FASTQ format can contain several sequences. FASTQ is a text-based format for storing both a biological sequence (usually nucleotide sequence) and its corresponding quality scores. It is mainly used for storing the output of high-throughput sequencing instruments. A FASTQ file usually uses four lines per sequence.

4. a '@' character, followed by a sequence identifier and an optional description
5. the raw sequence letters.
6. a '+' character, optionally followed by the same sequence identifier (and any description)
7. quality values for the sequence in Line 2

An Example Sequence in FASTQ Format is

@SEQUENCE_ID

G T G G A A G T T C T T A G G G C A T G G C A A A G A G T C A G A A T T T G A C

+

F A F F A D E D G D B G E G G B C G G H E > E E B A @ @ =

For example: -

>crab chick ALPHA CRYSTALLIN B CHAIN (ALPHA(B)-CRYSTALLIN).

M D I T I H N P L V R R P L F S W L T P S R I F D Q I F G E H L Q E S E L L P T S P S L S P F L M R

S P F F R M P S W L E T G L S E M R L E K D K F S V N L D V K H F S P E E L K V K V L G D M I E I H

G K H E E R Q D E H G F I A R E F S R K Y R I P A D V D P L T I T S S L S L D G V L T V S A P R K Q

S D V P E R S I P I T R E E K P A I A G S Q R K

Fasta

FASTA is a DNA and protein sequence alignment software package first described by David J. Lipman and William R. Pearson in 1985.[1] Its legacy is the FASTA format which is now ubiquitous in bioinformatics.

FASTA- fast All (Fast Alignment Search Tool)

The original FASTP program was designed for protein sequence similarity searching. Because of the exponentially expanding genetic information and the limited speed and memory of computers in the 1980s heuristic methods were introduced aligning a query sequence to entire data-bases. FASTA, published in 1987, added the ability to do DNA:DNA searches, translated protein: DNA searches, and also provided a more sophisticated shuffling program for evaluating statistical significance.[2] There are several programs in this package that allow the alignment of protein sequences and DNA sequences. Nowadays, increased computer performance makes it possible to perform searches for local alignment detection in a database using the Smith–Waterman algorithm.

FASTA is pronounced "fast A", and stands for "FAST-All", because it works with any alphabet, an extension of the original "FAST-P" (protein) and "FAST-N" (nucleotide) alignment tools.

What is FASTA Format?

Base pairs or amino acids are represented using single-letter codes in the text-based FASTA format, which can be used to represent nucleotide sequences or peptide sequences. In FASTA format, a sequence consists of lines of sequence data after a single line description. The greater-than (">") symbol in the first column separates the description line from the sequence data. It is advised that text lines be no more than 80 characters long. The sequence represents **genetic information**.

Biological deoxyribonucleic acid represents the information which directs the functions of an organism (Lipman, DJ; Pearson, WR (1985) FASTA is pronounced "fast A", and stands for "FAST-All", because it works with any alphabet, an extension of the original "FAST-P" (protein) and "FAST-N" (nucleotide) alignment tools.(Pearson, WR; Lipman, DJ (1988) The FASTA file format used as input for this software is now largely used by other sequence database search tools (such as BLAST) and sequence alignment programs (Clustal, T-Coffee, etc.).

FASTA format is a text-based format for representing either nucleotide sequences or amino acid (protein) sequences, in which nucleotides or amino acids are represented using single-letter codes.

In the original format, a sequence was represented as a series of lines, each of which was no longer than 120 characters and usually did not exceed 80 characters. This probably was to allow for the preallocation of fixed line sizes in software: at the time most users relied on Digital Equipment Corporation (DEC) VT220 (or compatible) terminals which could display ٨٠ or ١٣٢ characters per line. Most people preferred the bigger font in -٨٠character modes and so it became the recommended fashion to use ٨٠ characters or less (often ٧٠) in FASTA lines. Also, the width of a standard printed page is ٧٠ to ٨٠ characters (depending on the font). Hence, ٨٠ characters became the norm.

The first line in a FASTA file started either with a ">" (greater-than) symbol or, less frequently, a ";" (semicolon) was taken as a comment. Subsequent lines starting with a semicolon would be ignored by software. Since the only comment used was the first, it quickly became used to hold a summary description of the sequence, often starting with a unique library accession number, and with time it has become commonplace to always use ">" for the first line and to not use ";" comments (which would otherwise be ignored)

An Example Sequence in FASTA Format is

>gi|186681228|ref|YP_001864424.1| phycoerythrobilin: ferredoxin oxidoreductase

M N S E R S D V T L Y Q P F L D Y A I A Y M R S R L D L E P Y P I P T G F E S N S A V V G K G K N Q E E V V T T S Y A F Q T A K L R Q I R A

A H V Q G G N S L Q V L N F V I F P H L N Y D L P F F G A D L V T L P G G H L I A L D M Q P L F R D D S A Y Q A K Y T E P I L P I F H A H Q

Q H L S W G G D F P E E A Q P F F S P A F L W T R P Q E T A V V E T Q V F A A F K D Y L K A Y L D F V E Q A E A V T D S Q N L V A I K Q A Q

L R Y L R Y R A E K D P A R G M F K R F Y G A E W T E E Y I H G F L F D L E R K L T V V K

Sequence is expected to be represented in the standard IUB/IUPAC amino acid and nucleic acid codes, with these exceptions:

- lower-case letters are accepted and are mapped into upper-case;
- a single hyphen or dash can be used to represent a gap of indeterminate length;
- in amino acid sequences, U and * are acceptable letters (see below).
- any numerical digits in the query sequence should either be removed or replaced by appropriate letter codes (e.g., N for unknown nucleic acid residue or X for unknown amino acid residue).

The nucleic acid codes are:

A --> adenosine M --> A C (amino)

C --> cytidine S --> G C (strong)

G --> guanine W --> A T (weak)

T --> thymidine B --> G T C

U --> uridine D --> G A T

R --> G A (purine) H --> A C T

Y --> T C (pyrimidine) V --> G C A

K --> G T (keto) N --> A G C T (any)

— gap of indeterminate length

The accepted amino acid codes are

An ALA alanine P PRO proline

B ASX aspartate or asparagine Q GLN glutamine

C CYS cystine R ARG arginine

D ASP aspartate S SER serine

E GLU glutamate T THR threonine

F PHE phenylalanine U selenocysteine

G GLY glycine V VAL valine

H HIS histidine W TRP tryptophan

I ILE isoleucine Y TYR tyrosine

K LYS lysine Z GLX glutamate or glutamine

L LEU leucine X any

M MET methionine * translation stop

N ASN asparagineZ - gap of indeterminate length

FASTA format is a very widely used (and abused) format. It consists of a header line starting with a > character followed by a code identifying the sequence and, very often, some text describing the sequence. The header line is followed by one or more lines containing the sequence itself. FASTA files may contain one or more sequences

>pdb|8PFE|C Chain C, Neuropilin-M G S S H H H H H H G S G SG L E V L F Q G P H M F K C M K A L G M E S G K I H S K Q I T A S S Q Y S T N W S A E R S R L N Y P E N G W T P G E D S Y R E W I Q V D L G L L R F V T V G T Q G A I S K E T K K K Y Y V K T Y K I D V S S N G K D W I T I E E G N K P V L F Q G N T N P T D V V V A V F P E P L I T R F V R I K P A T W E T G I S M R F E V Y G C K I T

GENE BANK

GENE BANK SEQUENCE FORMAT

- The GenBank format was developed by the U.S. National Center for Biotechnology Information (NCBI). It is widely used by public databases and is considered by many to be the standard DNA and protein sequence file format. NOTE: GenBank sequence files also use the .GB file extension and more commonly, the. GBK extension.
- **GENBANK** stands for National Genetic Sequence Data Base. Suggest new definition. This definition appears very rarely and is found in the following Acronym Finder categories: Science, medicine, engineering, etc.

GenBank is a comprehensive public database of nucleotide sequences and supporting bibliographic and biological annotation. GenBank is built and distributed by the National Center for Biotechnology Information (NCBI), a division of the National Library of Medicine (NLM), located on the campus of the US National Institutes of Health (NIH) in Bethesda, USA.

FASTA format is a text-based format for representing either nucleotide sequences or amino acid (protein) sequences, in which nucleotides or amino acids are represented using single-letter codes. The format also allows for sequence names and comments to precede the sequences

CLUSTAL Format: Used to represent multiple sequence alignments.

Example: sequence1 ATGCT, sequence2 AT-C-

PHYLIP Format: Compact format for phylogenetic analysis with alignments.

Annotation Formats

GenBank Format: Contains sequences and rich annotation, such as gene location and functional information.

Example snippet: LOCUSSCU49845 5028 bp DNA linear PLN

ORIGIN

1 agcttagcta...

GFF (General Feature Format): Stores annotations for genomic sequences.

Example: chr1. gene 1300 9000. + . ID=gene1; Name=GeneA

1. Structure Formats: DB (Protein Data Bank) Format: Stores 3D structural data of biomolecules.

   Example snippet: HEADER HYDROLASE 01-JAN-00

ATOM 1 N MET A 1 58.380 50.475 27.57

MMTF (Macromolecular Transmission Format): Binary format optimized for molecular structures.

2. Specialized Formats' (Variant Call Format):
3. Stores genomic variations: Example: #CHROM POS ID REF ALT QUAL FILTER INFO

   20 14370 rs605 G A 29.0 PASS.

SAM/BAM Format

Represents sequence alignments. BAM is the binary equivalent of SAM.

Scope of Sequence Formats in Bioinformatics Work

1. Data Storage and Management: Sequence formats facilitate efficient storage of large-scale biological datasets, such as genome assemblies and transcriptomes.
2. Data Exchange: Standardized formats allow interoperability between bioinformatics tools, databases (e.g., NCBI, Ensembl), and researchers.
3. Sequence Analysis: Formats like FASTA and FASTQ are foundational for applications such as: Sequence alignment (e.g., BLAST, ClustalW).

Gene prediction and annotation

1. **Structural Biology:** Formats like PDB and MMTF enable the analysis of biomolecular structures for drug discovery and protein engineering.
2. **Genomics and Personalized Medicine:** Formats like VCF and GFF are critical for variant analysis, which supports personalized medicine initiatives.
3. **Machine Learning and Computational Analysis:** Modern bioinformatics pipelines utilize sequence formats as input for machine learning algorithms in tasks like protein structure prediction and mutation effect estimation.

Sequence Similarity

In bioinformatics, BLAST (Basic Local Alignment Search Tool) is an algorithm for comparing primary biological sequence information, such as the amino-acid sequences of different proteins or the nucleotides of DNA sequences. A BLAST search enables a researcher to compare a query sequence with a library or database of sequences, and identify library sequences that resemble the query sequence above a certain threshold. Different types of BLASTs are available according to the query sequences. For example, following the discovery of a previously unknown gene in the mouse, a scientist will typically perform a BLAST search of the human genome to see if humans carry a similar gene; BLAST will identify sequences in the human genome that resemble the mouse gene based on similarity of sequence.

Similarity in sequence alignment is the resemblance between two sequences when compared. This fact is dependent on the identity of sequences. Similarity depicts the extent to which the residues are aligned. Hence, similar sequences contain similar properties.

Basic Scoring Matrices (PAM AND BLOSUM)

Scoring matrices are used to determine the relative **score** made by matching two characters in a sequence alignment. There are many flavors of **scoring matrices** for amino acid sequences, nucleotide sequences, and codon sequences, and each is derived from the alignment of "known" homologous sequences.

The **PAM** matrices are based on scoring all amino acid positions in related sequences, whereas the **BLOSUM** matrices are based on substitutions and conserved positions in blocks, which represent the most-alike common regions in related sequences.

A **PAM** matrix is a matrix where each column and row represents one of the twenty standard amino acids. In **bioinformatics**, **PAM** matrices are regularly used as substitution matrices to score sequence alignments for proteins.

Point accepted mutation — also known as a PAM — is the replacement of a single amino acid in the primary structure of a protein with another single amino acid, which is accepted by the processes of natural selection. This definition does not include all point mutations in the DNA of an organism.

In bioinformatics, the BLOSUM (BLOCKS Substitution Matrix) matrix is a substitution matrix used for sequence alignment of proteins. BLOSUM matrices are used to score alignments between evolutionarily divergent protein sequences. They are based on local alignments.

Substitution matrices are used to score aligned positions in a sequence alignment procedure, usually of amino acids or nucleotide sequences.

Two Commonly Used Matrices

PAM and BLOSUM.

PAM = Percent Accepted Mutations (Margaret Dayhoff)

BLOSUM = Blocks Substitution Matrix (Steven and Henikoff)

BLOSUM matrices are used to score alignments between evolutionarily divergent protein sequences. They are based on local alignments. BLOSUM matrices were first introduced in a paper by Steven Henikoff and Jorja Henikoff.

Novel sequencing technologies like Illumina/Solexa, AB/SOLiD, and Roche/454 (Mardis, 2008) have led to the development of a number of new alignment tools that can effectively perform read mapping against large reference sequences, such as the human genome (Langmead et al., 2009; Li et al., 2008). However, the various formats in which these programmed generate alignments complicate downstream processing. A clear interface is created between alignment and subsequent studies, including as variant detection, genotyping, and assembly, by using a standard alignment format that is compatible with all sequence types and aligners.

Importance of Sequence Format

The sequence format is critical in bioinformatics because it determines how data is read, processed, and interpreted by various computational tools. Here's an overview of its importance, with references:

Compatibility

Different bioinformatics tools require specific sequence formats (e.g., FASTA, GenBank, PDB). For example:

I.FASTA is commonly used for sequence alignment tools like BLAST.

II. GenBank includes annotation data needed for genome analysis.

Future sequence formats in bioinformatics are expected to evolve to address the increasing complexity and scale of biological data. These advancements aim to enhance data storage, interoperability, annotation detail, and computational efficiency. Here are some anticipated developments:

Scalable Formats for Large Datasets

With the rise of high-throughput sequencing technologies (e.g., long-read sequencing by PacBio or Oxford Nanopore), future formats will prioritize: Compression: Efficient storage for terabyte- or petabyte-scale datasets. Streaming capabilities: Enabling real-time processing of sequence data (Bonfield, J. K., et al. (2013)

Examples: CRAM (Compressed Reference-Aligned BAM): Already used for efficient storage of aligned sequences, with potential for further optimization.

Standardization for Multi-Omics Integration

Future formats will likely combine genomics, transcriptomics, proteomics, and metabolomics data, ensuring a unified structure for holistic analyses. These formats will: Allow integration of diverse biological data types. Include metadata, annotations, and functional interactions in a cohesive framework.

Example: The development of multi-layered formats like HDF5 (Hierarchical Data Format), used in single-cell RNA sequencing (e.g., Seurat).

Interactive and Machine-Readable Formats

Future formats will be designed for machine learning and AI integration: Graph-based formats: Represent complex genomic variations (e.g., SVs, CNVs) more effectively than linear formats. Examples: Graph Genomes, which visualize population-scale genomic diversity. Interactivity: Formats enabling real-time annotation, visualization, and edits.

User-Friendly and Plug-and-Play Formats

To lower entry barriers for researchers: Formats will include automated validation checks. And integrate with visualization tools (e.g., genome browsers).

Conclusion

To sum up, sequence formats are critical to the storage and analysis of biological sequence data in bioinformatics, including DNA, RNA, and protein sequences. These formats have distinct functions and are employed in a range of bioinformatics research applications: FASTA: extensively employed to store sequences and their unique IDs. frequently employed for database searches, sequence retrieval, and alignment of sequences. GenBank show in Detailed nucleotide sequence format with annotations and metadata. used for database searches, data exchange, and genetic annotation. FASTQ role saves quality scores and sequencing data. vital for read alignment, quality control, and analysis of next-generation sequencing. Choosing the correct sequence format is crucial for ensuring accuracy, compatibility, and reproducibility in bioinformatics workflows. Always adhere to the required format guidelines of the tool or database you're working with. Future sequence formats will balance the needs of scalability, integration, security, and ease of use to support the next generation of bioinformatics research.

Emerging standards like graph-based and cloud-compatible formats will play a pivotal role in addressing modern challenge

Refrences

1. Lipman DJ, Pearson WR. "Rapid and sensitive protein similarity searches." Science, 1985.
2. Benson DA, Karsch-Mizrachi I, Lipman DJ, et al. "GenBank." Nucleic Acids Research, 2019.
3. Thompson JD, Higgins DG, Gibson TJ. "CLUSTAL W: improving the sensitivity of progressive multiple sequence alignment through sequence weighting, position-specific gap penalties and weight matrix choice." Nucleic Acids Research, 1994.
4. Berman HM, Westbrook J, Feng Z, et al. "The Protein Data Bank." Nucleic Acids Research, 2000.
5. Lipman, DJ; Pearson, WR (1985). "Rapid and sensitive protein similarity searches". Science. **227** (4693):,1435,41. Bibcode:1985Sci...227.1435L. doi:10.1126/science.2983426. PMID 2983426.
6. Pearson, WR; Lipman, DJ (1988). "Improved tools for biological sequence comparison". Proceedings of the National Academy of Sciences of the United States of America. 85 (8):2444,8. Bibcode:1988PNAS...85.2444P. doi:10.1073/pnas.85.8.2444. PMC 280013. PMID 3162770.
7. David W. Mount: Bioinformatics Sequence and Genome Analysis, Edition 1, Cold Spring Harbor Laboratory Press, 2001, pp. 295–297.
8. Cock, P.J. et al. (2010). "The Sanger FASTQ file format for sequences with quality scores." Nucleic Acids Research.
9. Cock, P. J. A., et al. (2009). "Bio python: freely available Python tools for computational molecular biology and bioinformatics." Bioinformatics, 25(11), 1422-1423. https://doi.org/10.1093/bioinformatics/btp163
10. Erlich, Y., & Narayanan, A. (2014). "Routes for breaching and protecting genetic privacy." Nature Reviews Genetics. https://doi.org/10.1038/nrg3723
11. Altschul, S.F., Gish, W., Miller, W., Myers, E.W. and Lipman. (1990) Basic local alignment search tool. J. Mol. Biol., 215, 403–410.
12. http://emboss.open-bio.org/html/use/ch05s02.html
13. Langmead B, et al. Ultrafast and memory-efficient alignment of short DNA sequences to the human genome, Genome Biol.2009, vol. 10 pg. R25
14. Mardis ER, Next-generation DNA sequencing methods, Annu. Rev. Genomics Hum. Genet., 2008, vol. 9 (pg. 387-402)
15. Benson, D.A. et al. (2013). "GenBank." Nucleic Acids Research.
16. Pertea, M., & Salzberg, S.L. (2010). "Using GFF files to store annotation data." Bioinformatics.
17. Li, H., & Durbin, R. (2009). "Fast and accurate short read alignment with Burrows–Wheeler transform." Bioinformatics.
18. Bonfield, J. K., et al. (2013). "The CRAM format—compressed reference-oriented alignment format." Bioinformatics. https://doi.org/10.1093/bioinformatics/btt330

3 Sequence Alignments

Introduction

The nucleotide sequence is considered the blueprint of an organism as it contains the instructions for the building of an organism. Similarly, the amino acid sequence provides important biochemical information about the protein's 3-D structure, function and evolution.

Sequence alignment is a basic process which compares two or more sequence, be it DNA, RNA or proteins. Using this process, one tries to understand which sites within the sequence are homologous or conserved and thus share a common evolutionary history. Comparison of amino acid sequences between different organisms provides insight to the function and conserveness of the protein. Considering the amino acid sequence, if the sequence is exactly same in two different organisms, it indicates that the proteins are same and thus have the same function. There are various software and tools which can be used to align various nucleic acid or protein sequences. Before we start with sequence alignment, we should be aware of the concepts of sequence similarity, identity and homology. Ideally, these three terms appear almost same but when we consider the alignment of nucleotides or amino acids; similarity, identity and homology have their importance.For alignment, the sequences are taken from their respective databases and should be in FASTA format.

Fasta Format

FASTA format is the representation of the nucleotide or peptide sequence in a text-based manner, where the respective nucleotides or amino acids are represented with their single letter code. A single line description starting with a greater than sign (>) precedes the sequence. The example of FASTA format of protein albumin from humans can be seen in **Figure 1**.

albumin [Homo sapiens]

GenBank: AAA98797.1

GenPept Identical Proteins Graphics

```
>AAA98797.1 albumin [Homo sapiens]
MKWVTFISLLFLFSSAYSRGVFRRDAHKSEVAHRFKDLGEENFKALVLIAFAQYLQQCPFEDHVKLVNEV
TEFAKTCVADESAENCDKSLHTLFGDKLCTVATLRETYGEMADCCAKQEPERNECFLQHKDDNPNLPRLV
RPEVDVMCTAFHDNEETFLKKYLYEIARRHPYFYAPELLFFAKRYKAAFTECCQAADKAACLLPKLDELR
DEGKASSAKQRLKCASLQKFGERAFKAWAVARLSQRFPKAEFAEVSKLVTDLTKVHTECCHGDLLECADD
RADLAKYICENQDSISSKLKECCEKPLLEKSHCIAEVENDEMPADLPSLAADFVESKDVCKNYAEAKDVF
LGMFLYEYARRHPDYSVVLLLRLAKTYETTLEKCCAAADPHECYAKVFDEFKPLVEEPQNLIKQNCELFE
QLGEYKFQNALLVRYTKKVPQVSTPTLVEVSRNLGKVGSKCCKHPEAKRMPCAEDYLSVVLNQLCVLHEK
TPVSDRVTKCCTESLVNRRPCFSALEVDETYVPKEFNAETFTFHADICTLSEKERQIKKQTALVELVKHK
PKATKEQLKAVMDDFAAFVEKCCKADDKETCFAEEGKKLVAASQAALGL
```

Figure 1: FASTA format of Albumin of Homo sapiens retrieved from GenBank

In the FASTA sequence, we can see the name of the protein, organism and GenBank Id at the top of the sequence. The sequence is preceded with the '>' greater than sign, GenBank Id, protein name and the scientific name of the organism.

Sequence Alignment and Sequence Similarity

Two sequences are considered identical when the nucleotide or amino acids exactly matches between the two sequences. For example, actin amino acid sequence from Gallus gallus and Acipenser ruthenuswere compared, as shown in Figure 2.

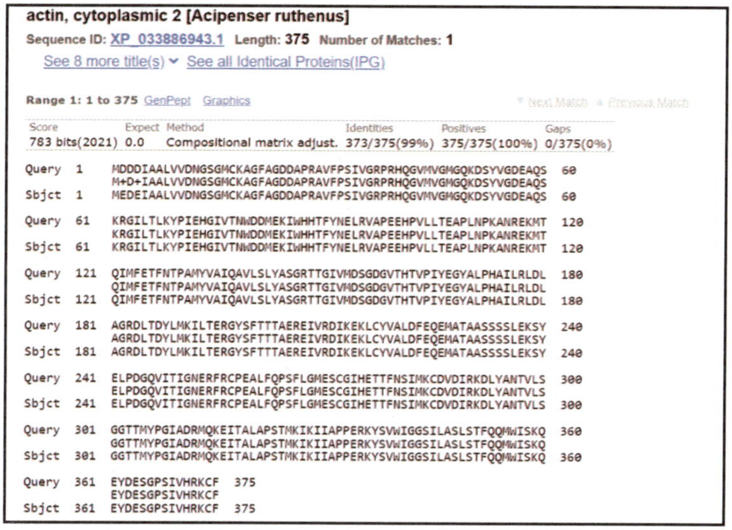

actin, cytoplasmic 2 [Acipenser ruthenus]

Sequence ID: XP_033886943.1 Length: 375 Number of Matches: 1

See 8 more title(s) See all Identical Proteins(IPG)

Range 1: 1 to 375 GenPept Graphics ▼ Next Match ▲ Previous Match

Score	Expect	Method	Identities	Positives	Gaps
783 bits(2021)	0.0	Compositional matrix adjust.	373/375(99%)	375/375(100%)	0/375(0%)

```
Query  1    MDDDIAALVVDNGSGMCKAGFAGDDAPRAVFPSIVGRPRHQGVMVGMGQKDSYVGDEAQS  60
            M+D+IAALVVDNGSGMCKAGFAGDDAPRAVFPSIVGRPRHQGVMVGMGQKDSYVGDEAQS
Sbjct  1    MEDEIAALVVDNGSGMCKAGFAGDDAPRAVFPSIVGRPRHQGVMVGMGQKDSYVGDEAQS  60

Query  61   KRGILTLKYPIEHGIVTNWDDMEKIWHHTFYNELRVAPEEHPVLLTEAPLNPKANREKMT  120
            KRGILTLKYPIEHGIVTNWDDMEKIWHHTFYNELRVAPEEHPVLLTEAPLNPKANREKMT
Sbjct  61   KRGILTLKYPIEHGIVTNWDDMEKIWHHTFYNELRVAPEEHPVLLTEAPLNPKANREKMT  120

Query  121  QIMFETFNTPAMYVAIQAVLSLYASGRTTGIVMDSGDGVTHTVPIYEGYALPHAILRLDL  180
            QIMFETFNTPAMYVAIQAVLSLYASGRTTGIVMDSGDGVTHTVPIYEGYALPHAILRLDL
Sbjct  121  QIMFETFNTPAMYVAIQAVLSLYASGRTTGIVMDSGDGVTHTVPIYEGYALPHAILRLDL  180

Query  181  AGRDLTDYLMKILTERGYSFTTTAEREIVRDIKEKLCYVALDFEQEMATAASSSSLEKSY  240
            AGRDLTDYLMKILTERGYSFTTTAEREIVRDIKEKLCYVALDFEQEMATAASSSSLEKSY
Sbjct  181  AGRDLTDYLMKILTERGYSFTTTAEREIVRDIKEKLCYVALDFEQEMATAASSSSLEKSY  240

Query  241  ELPDGQVITIGNERFRCPEALFQPSFLGMESCGIHETTFNSIMKCDVDIRKDLYANTVLS  300
            ELPDGQVITIGNERFRCPEALFQPSFLGMESCGIHETTFNSIMKCDVDIRKDLYANTVLS
Sbjct  241  ELPDGQVITIGNERFRCPEALFQPSFLGMESCGIHETTFNSIMKCDVDIRKDLYANTVLS  300

Query  301  GGTTMYPGIADRMQKEITALAPSTMKIKIIAPPERKYSVWIGGSILASLSTFQQMWISKQ  360
            GGTTMYPGIADRMQKEITALAPSTMKIKIIAPPERKYSVWIGGSILASLSTFQQMWISKQ
Sbjct  301  GGTTMYPGIADRMQKEITALAPSTMKIKIIAPPERKYSVWIGGSILASLSTFQQMWISKQ  360

Query  361  EYDESGPSIVHRKCF  375
            EYDESGPSIVHRKCF
Sbjct  361  EYDESGPSIVHRKCF  375
```

Figure 2: Sequence identity between actin protein from Gallus gallus and Acipenser *ruthenus*.

The sequences presented in the red box are exactly same and thus represent the identity but the amino acid represented in green box are different but are acidic amino acids, i.e., aspartic acid (D) and glutamic acid (D). As it can clearly be seen that the identity between the two sequences is 99%.

On the contrary, sequence similarity represents the resemblance between the two nucleotide or amino acid sequences. Consider, amino acids represented in green box in Figure 2 we can see that the two sequence have different amino acids and thus not identical but both the amino acids are acidic, having the same physicochemical property and thus are similar. If it would have been an acidic amino acid and a basic amino acid, such as D (Aspartic acid) and L (Lysine), respectively then we would have easily said that at that particular position the sequence is neither identical nor similar.

Similarly in the case of nucleotide, adenine (A) and guanine (G) are purines and thymine (T) and cytosine (C) are pyrimidines suggesting that if in one sequence A and in another sequence, G is present at a specific position, then it represents sequence similarity as being the purines they have similar physicochemical properties.

Types of Sequence Alignment

The sequence alignment is further categorized as global and local alignment. In case of global alignment, the sequencesare aligned end to end whereas in case of local alignment, short stretches of sequences with highest matches are aligned (Figure 3).

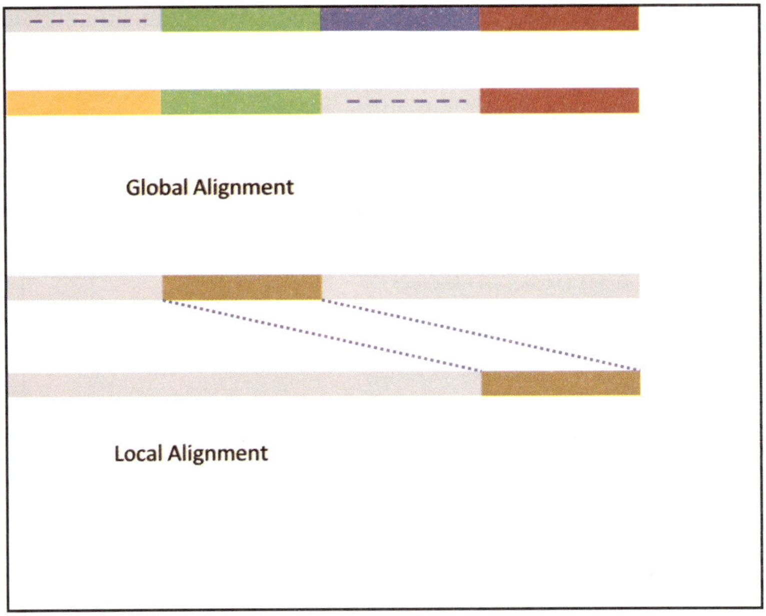

Figure 3: Pictorial representation of global and local alignment. As observed in the image, global alignment whole sequence is aligned whereas in local alignment short stretches of sequence represented as brown block are aligned.

Further sequence alignment is also categorized as pairwise alignment and multiple sequence alignment depending on the number of sequences aligned. Pairwise sequence alignment involves identifying regions of similarity between two sequences which indicate the structural, functional or evolutionary relation between the two sequences. Pairwise alignment can be done using either local or global alignment. Multiple sequence involves identifying regions of similarity between more than two sequences which indicates the functional, structural or evolutionary relationship between aligned sequences.

Pairwise alignment

When comparing two nucleotide sequences or amino acid sequences identification of regions of similarity or identity indicates functional, structural or evolutionary relationship among the species. Figure 4represents the pairwise alignment between two amino acid sequences. It can be clearly seen that out of nine amino acids, seven of the amino acids are identical and aligned.

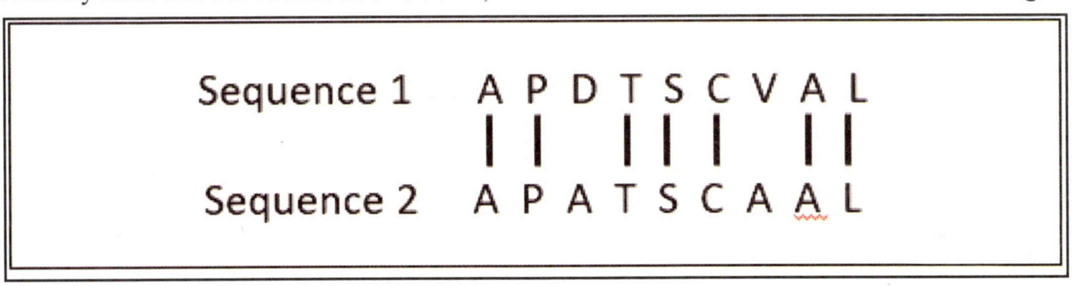

Figure4: Representative example of pairwise alignment between two amino acid sequences.

There are different methods for analysing the alignment between two sequence which includes:

19. Dot Plot: The pairwise alignment can be graphically represented in the form of dot plot.

Beside representing the identity or similarity, dot plots also exhibit repetitions including tandem repeats, inversions, palindromic regionsand deletions can be represented by the dot plots (Cabanettes et al., 2018). In this plot, the sequences are mentioned horizontally and vertically. Each amino acid or nucleotides are represented as single letter code. If one amino acid or nucleotide matches with another amino acid or nucleotide, then a dot is kept in the respective boxes. Figure 5indicates a representative dot plot and the dot plot between actin of Drosophila melanogaster (NP_001284915.1) and Gallus gallus (NP_990849.1). EMBOSS Dotmatcher tool was used to perform the dot plot (Madeira et al., 2022). The continuous main diagonal indicates a perfect similarity between the sequences (Figure 5 b). The parallels to the main diagonal represent the repeated regions in the same direction on different parts of the sequence (represented by red circles in figure 5 b).

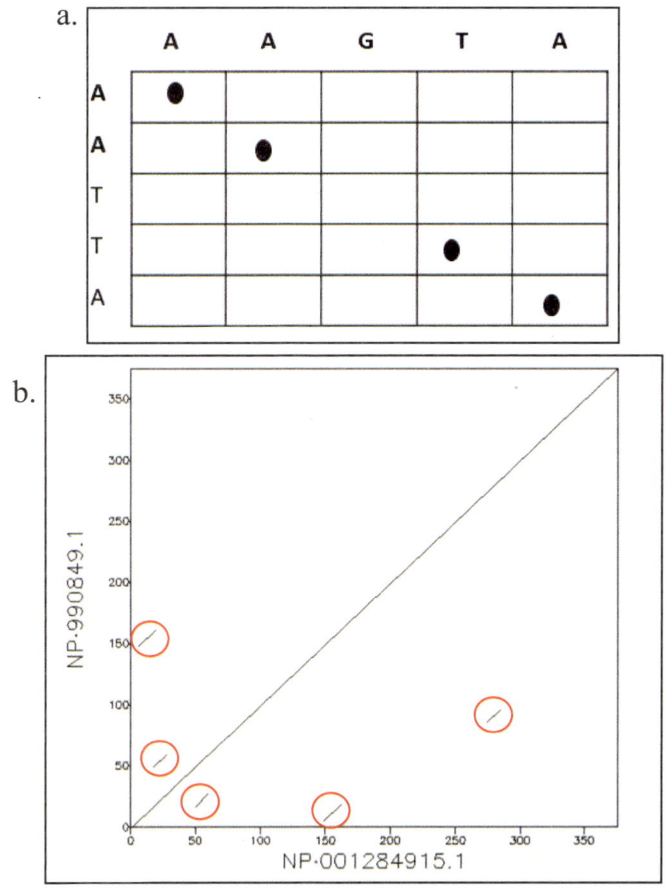

Figure 5: (a)Representative image of dot plot (b) Dot plot between actin of Drosophila melanogaster and Gallus gallus indicates identity between the 2 proteins indicating same function.

Dynamic Programming is used for optimal alignment of the sequences. Here, the sequences are broken into sub-parts which are further solved to compute the solution of the larger one. In dynamic programming cost benefit approach is considered for quantitative alignment (Arslan et al., 2004). Needleman-Wunsch and Smith-Waterman algorithms use dynamic programming. Three scores are specified in this case: 1. Positive score for alignment of two characters (nucleotide or amino acids) in both the sequences; 2. Zero or negative score for aligning a pair of sites containing different characters (or mismatch) in the sequences; 3. Zero or negative score for aligning a pair of site in one sequence with gap in another sequence (or gap). The score given for match and mismatch is referred to as penalty. Consider the example shown in figure 6 which represent the scoring for two sequences.

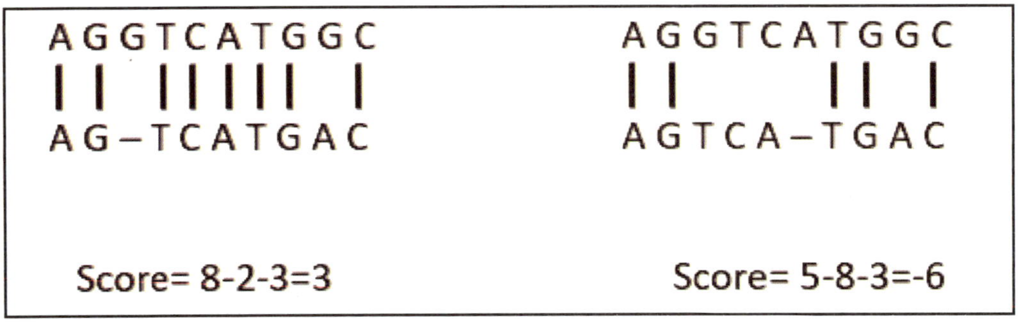

Figure 6: Alternate alignment of a pair of sequences indicating a simple scoring function with matches=+1, mismatches=-2 and gap=-3. The alignment with the score 3 is better than the alignment with the score -6 because of the large score.

Heuristic method is used when we have to compare a single sequence against the complete database such as BLAST (Altschul et al., 1990). This method gives an approximate solution to the given problem. This method is used as it is faster than dynamic programming.

EMBOSS Needle is on of the bioinformatics tool which is used for pairwise alignment. Here both protein and DNA can be used as input sequence **(Madeira et al., 2022)**. The hemoglobin protein sequence from Homo sapiens and Mycobacterium tuberculosis NCGM2209. The alignment result is shown in the figure 7.

```
# Aligned_sequences: 2
# 1: EMBOSS_001
# 2: EMBOSS_001
# Matrix: EBLOSUM62
# Gap_penalty: 10.0
# Extend_penalty: 0.5
#
# Length: 188
# Identity:      25/188 (13.3%)
# Similarity:    40/188 (21.3%)
# Gaps:         104/188 (55.3%)
# Score: 27.5
#
#
#=======================================

EMBOSS_001         1 ----------------------------------MVLSPADKTNVKAAWG     16
                                                       .:.:....||:....|
EMBOSS_001         1 MRKREPISIYDKIGGHEAIEVVVEDFYVRVLADDQLSAFFSGTNMSRLKG     50

EMBOSS_001        17 KVGAHAGEYGAEAL--ERMFLSFPTTKTY------FPHFDLSHGSAQVKG     58
                     |    ..|:.|.||  ...:...|..:.:      ..||.|      |.|
EMBOSS_001        51 K----QVEFFAAALGGPEPYTGAPMKQVHQGRGITMHHFSL------VAG     90

EMBOSS_001        59 HGKKVADALTNAVAHVDDMPNALSALS----DLHAHKLRVDPVNFKLLSH    104
                     |   :|||||.|....:.:...|..::    |:.:.:....||
EMBOSS_001        91 H---LADALTAAGVPSETITEILGVIAPLAVDVTSGESTTAPV-------    130

EMBOSS_001       105 CLLVTLAAHLPAEFTPAVHASLDKFLASVSTVLTSKYR     142

EMBOSS_001       131 --------------------------------------     130
```

Figure 7: Pairwise alignment using EMBOSS Needle using default setting indicates a score of 27.5 indicating the two proteins are not well aligned and thus are not structurally similar. Symbol 'I' represents identical amino acids, ':' represents amino acids with similar properties and '.' represents different amino acids.

Multiple Sequence Alignment

Multiple sequence alignment is an important tool for structure and function prediction of proteins, phylogenetic analysis and other jobs related to sequence analysis (Edgar et al., 2006). CLUSTAL omega is the frequently used multiple sequence alignment tool but there are other tools which are used for better results, these are PREFAB, SABMARK, OXBENCH and IRMBASE (Edgar et al., 2006).

Consider the multiple sequence alignment of hemoglobin gene from 5 different organisms using clustal omega (Figure 8). On the result menubar if we click on guid tree we can observe the phylogram of all the organisms for which multiple sequence alignment was performed. Phylogram is the phylogenetic tree where the branch length is directly proportional to the branch length (Figure 9).

```
BAN62839.1                 MSAFNPHEKQLIGDLWHKVDVAHCGGEALSRMLIVYPWKRRYFENFGDISNAQAIIHNEK 60
sp|P68871.2|HBB_HUMAN      MVHLTPEEKSAVTALWGKVNVDEVGGEALGRLLVVYPWTQRFFESFGDLSTPDAVMGNPK 60
XP_054296805.1             MVHLTPEEKSAVTALWGKVNVDEVGGEALGRLLVVYPWTQRFFESFGDLSTPDAVMGNPK 60
NP_990820.1                MVHWTAEEKQLITGLWGKVNVAECGAEALARLLIVYPWTQRFFASFGNLSSPTAILGNPM 60
pdb|1HV4|H                 -VHWSAEEKQLITGLWGKVNVADCGAEALARLLIVYPWTQRFFSSFGNLSSPTAILGNPM 59
                               . .**. :  ** **:* . *.***.*:*:****.:*:* .**::*.  *:: *

BAN62839.1                 VQAHGKKVLASFGEAVKHLDSIRAHFANLSKLHCEKFHVDPENFKLLGDIIIIVLAAHYP 120
sp|P68871.2|HBB_HUMAN      VKAHGKKVLGAFSDGLAHLDNLKGTFATLSELHCDKLHVDPENFRLLGNVLVCVLAHHFG 120
XP_054296805.1             VKAHGKKVLGAFSDGLAHLDNLKGTFAKLSELHCDKLHVDPENFRLLGNVLVCVLAHHFG 120
NP_990820.1                VRAHGKKVLTSFGDAVKNLDNIKNTFSQLSELHCDKLHVDPENFRLLGDILIIVLAAHFS 120
pdb|1HV4|H                 VRAHGKKVLTSFGDAVKNLDNIKNTFAQLSELHCDKLHVDPENFRLLGDILIIVLAAHFA 119
                           *.******* .*.:.: :**.::  *: **:***:*:*******:***:::: *** *:

BAN62839.1                 KDFGLECHAAYQKLVRQVAAALAAEYH  147
sp|P68871.2|HBB_HUMAN      KEFTPPVQAAYQKVVAGVANALAHKYH  147
XP_054296805.1             KEFTPQVQAAYQKVVAGVANALAHKYH  147
NP_990820.1                KDFTPECQAAWQKLVRVVAHALARKYH  147
pdb|1HV4|H                 KEFTPDCQAAWQKLVRVVAHALARKYH  146
                           *:*    :**:**:*  ** *** :**
```

Figure 8: Multiple sequence alignment of beta hemoglobin protein from four different organisms including Homo sapiens, Gallus gallus , Pongo pygmaeus, Crocodylus siamensis and Anser indicus. * (asterisk) indicates identical amino acids in all the four organisms, : (colon) represents amino acids with similar properties whereas'.' (dot) represents amino acids with weak similarity.

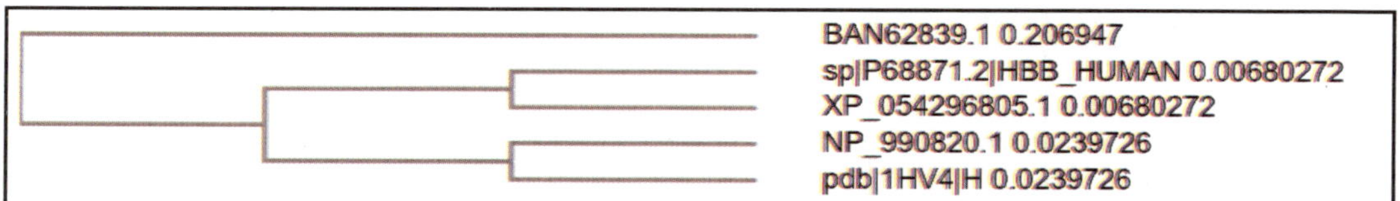

Figure 9: Phylogram of beta hemoglobin protein from four different organisms including Homo sapiens, Gallus gallus , Pongo pygmaeus, Crocodylus siamensis and Anser indicus.

Basic Local Alignment and Search Tools (Blast)

In order to determine the sequence similarity Basic local alignment search tool (BLAST) is the most frequently used tool. BLAST was developed by Steven Altschul and Samuel Karlin in 1990 (Altschulet al., 1990). BLAST uses a set of algorithms that identifies shorth stretch of query sequence that perfectly aligns with the fragment of the subject sequence found in the database.

The first step in BLAST is to break the sequence into short stretches or words of specific length. These words are series of nucleotides or amino acids from the query sequence. The default length is three characters during the search for protein sequences (Lobo, 2008) and for nucleic acid it is 28 nucleotides. For example, consider the 12 amino acid sequence near the amino terminal of actin from Gallus gallus

MDDIAALVVDNS

This sequence is then broken down into small words by selecting first amino acid, then moving to the next character, selecting the next three characters and so on to form the following words:

MDD DDI DIA IAA AAL ALV LVV VVD VDN DNS

During BLAST seed (identification of the word match), the algorithm finds all the common words between the query sequence and the subject or hit sequence. The regions with word match will be used to create an alignment (Figure 10).BLAST will be initiated by making words for the whole query sequence. The alignment should be greater than the neighbourhood score threshold (T). The alignment is increased in both the direction until unless T score does not continue to increase. The T score is derived using the scoring matrix. For protein searches, BLOSUM62 is the default scoring matrix.

Further there are different types of sequences, such as nucleotides and proteins and thus there are different BLAST programs depending on the type of sequences. BLAST program has five types: blastn, blastp, blastx, tblastn and tblastx (Summarized in Table 1).

The most commonly performed BLAST programs are blastn and blastp where nucleotides are compared with nucleotides and proteins are compared with proteins, respectively. Other BLAST program, such as tblastn where nucleotide query sequence is first translated and then results in protein hits. Thus, we can observe protein sequence of the sequenced gene.

Table 1: Summary of the different types of BLAST

BLAST type	Query	Database	Remarks	Reference
blastn	Nucleotide	Nucleotide	Compares nucleotide with nucleotide database	Boratyn et al., 2013
blastp	Protein	Protein	Compares proteins with protein database	Altschul et al., 1990
tblastn	Protein	Nucleotide	The database is translated into protein	Gertz et al., 2006
blastx	Nucleotide	Protein	The query is translated into protein	Gish and States, 1993 doi:10.1038/ng0393-266
tblastx	Nucleotide	Nucleotide	Both the query and the database are translated into the protein	Altschul et al., 1990

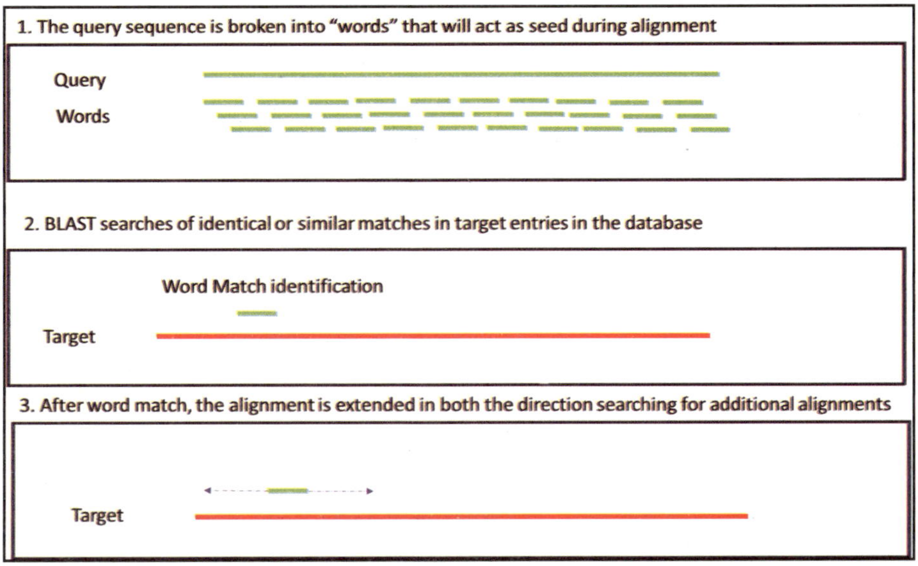

BLAST or Basic Local alignment search tool uses BLOSUM62 scoring Matrix as default. BLOSUM refers to Block Substitution Matrix where the sequence is broken down into segments or blocks (Henikoff and henikoff, 1992). The blocks are compared, where the identities are given as positive score, mismatch is given negative score or zero, gap initiation is also given a zero or negative score and highly negative score is given to gap extension. The numbers written after BLOSUM for e.g., BLOSUMx represents the comparison of sequences with no less than x% divergence. This suggest that BLOSUM62 compares sequences with less than x% divergence.

Besides BLOSUM62 there are other substituition matrices which beside matching the sequences consideres the structure of the protein as in some cases, proteins might be different but the protein structure or folds might be same. These substitution matrix are ProtSub and VTML200 (Jia and Jernigian, 2021)

Let us take an example of FtsZ protein from E.coli and run blast using the protein. Open NCBI, in the drop-down menu, click protein and type the name of the protein. Select the option of your choice and open the file and select the FASTA option (Figure 11).

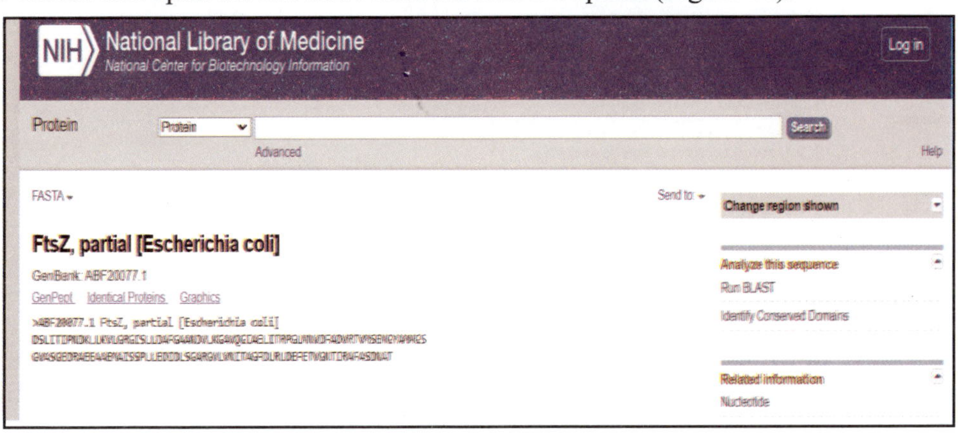

Figure 11: Protein FASTA file of E. coliFtsZ from NCBI

In order to identify proteins which are similar to FtsZ, we will run blastp. At the top we can see the blast options indicated by red circle. Enter the query sequence in the space provided indicated with orange circle (Figure 12).

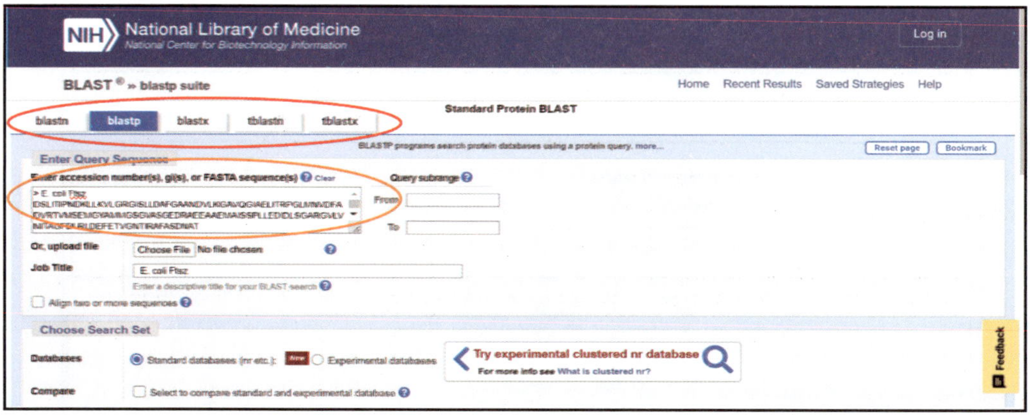

Figure 12: BLAST homepage. Red circle indicates the different BLAST parameters available and the orange circle indicates the place where query sequence needs to be added in FASTA format. Below this there is a check box 'align two or more sequence' by selecting it one can input the subject sequence as well. The result window will show alignment between the query and the selected subject sequence.

Further below after entering the query sequence, one can choose the database from where identical or similar sequences can be searched. By default, the database selected is 'non-redundant protein sequence (nr)', but based on the requirement one can choose any other database from the drop-down menu. For present example we have chosen 'protein data bank proteins (pdb)' as our search database (Figure 13).

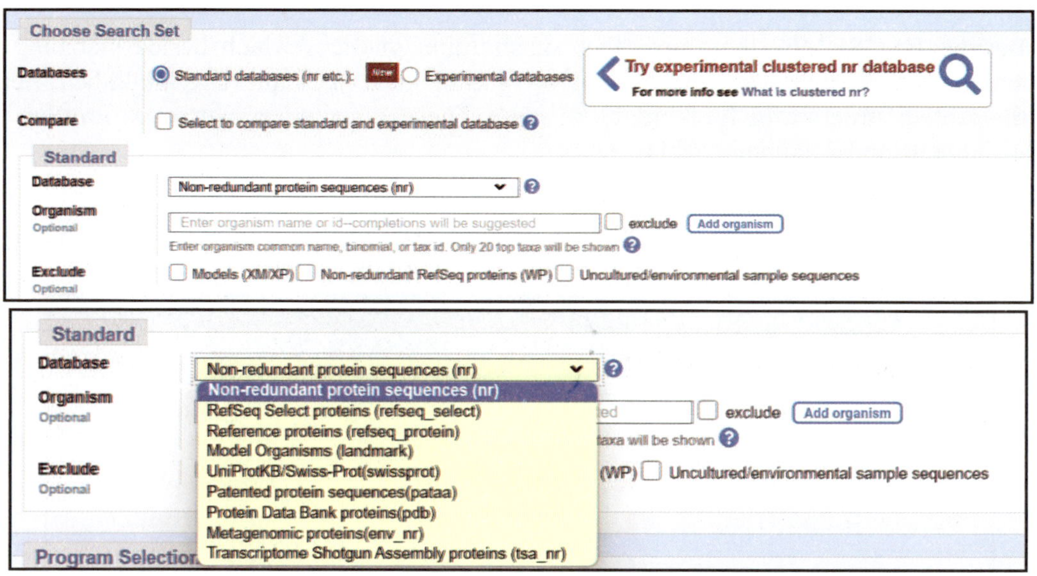

Figure 13: Databases available for searching for alignment for BLASTP.

As we move down the blast page, general parameters and scoring parameters are given. Default parameters are already selected but one can also change it based on their requirement. As we can observe, the maximum target sequences that will be shown is 100, the threshold value is kept at 0.05. The word size is 5 which is used as a seed for BLAST. The matrix being used is BLOSUM62, one can change it based on their result or analysis. In this case the gap cost for existence is -11 and for extension is -1. The more the gap lower will be the score indicating a bad alignment (Figure 14). Once the parameters are finalized, click on BLAST to start the alignment process.

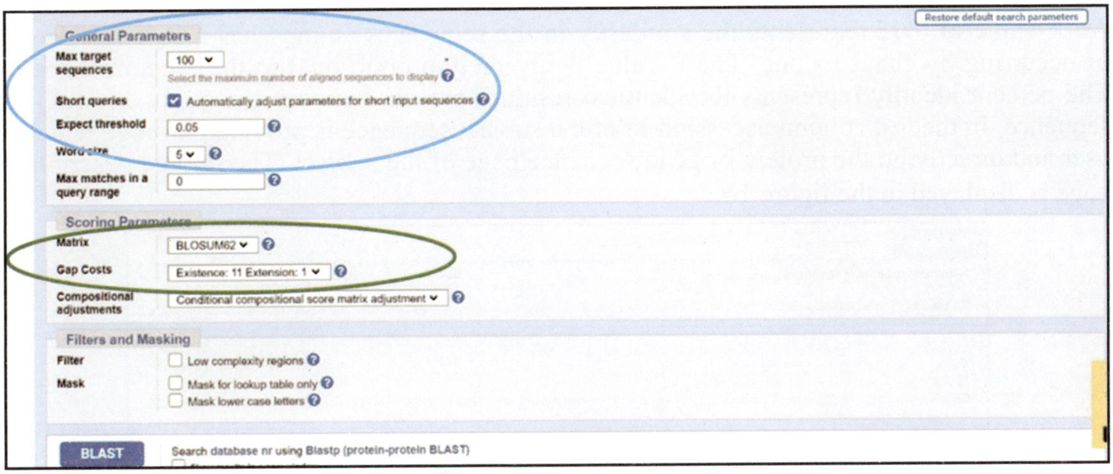

Figure 14: The general parameters indicating the word size, expect value and max target sequences are highlighted in blue circle. The Scoring parameters, such as matrix and gap cost are represented with green circle.

The window shown in figure 15 is the result window. The result window represents the job title *(E. coli Ftsz)*, program (BLASTP), the database against which the query was searched along with the query length is mentioned. The results of alignment are represented in various forms. The graphical summary represents the pictorial image of the alignment. The description shows the target no, query coverage, alignment percent, E-value and the accession number. The alignment tab shows the alignment with matches, mismatches, and gap. One can select the best alignment based on the score.

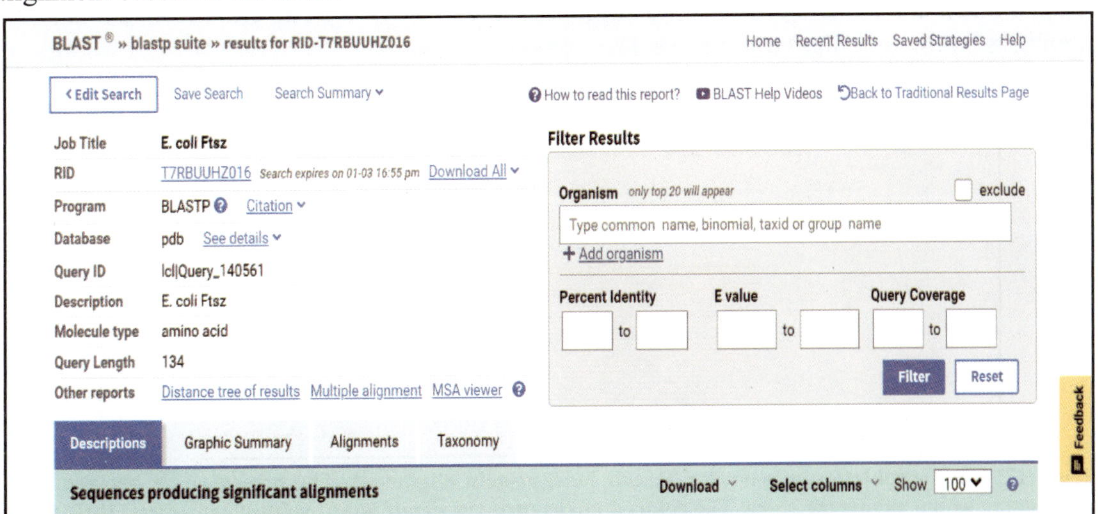

Figure 15: The result window of BLASTP represents the basic information of the program being run.

In the description, the information regarding the alignment is tabulated. The first row represents the description of the subject and second row gives the scientific of the organism from which the subject sequence has been identified. It is followed by the score of the alignment. Higher the score better is the alignment. Query coverage reference to the length of the query sequence which was aligned with the subject sequence. In the figure 16, we can observe that the query coverage ranges between 100-94%. The E-vale also referred to as expected value. An E-value of 0.05 represents the similarity in the sequences aligned has 5 in 100 chances of occurring by chance alone. The E-value is inversely proportional to the alignment score. The percent identity represents the identical residues present between the query and subject sequence. In the last column accession id of the subject sequence is provided through which a user and directly go the protein or gene sequence page of the subject. The sample description page is displayed in the figure 16.

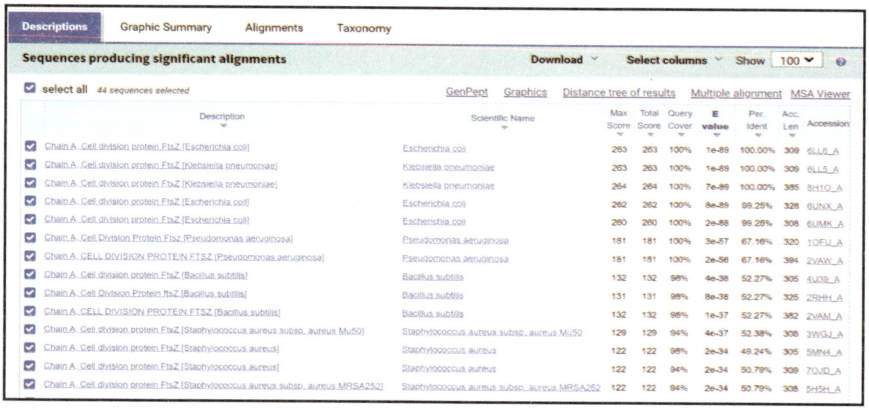

Figure 16: Description section obtained after BLAST run for E. coli FtsZ protein.

The BLAST output page also represents the result summary in graphical form where different lines with specific colors are present. Each line represents a different subject sequence and its length determines the query coverage. The color on the contrary represents the score of the alignment, for instance red color line represent an alignment score of more than or equal to 200 and black color line represents the alignment score below 40. The graphical summary for E. coli FtsZ protein is shown in Figure 17.

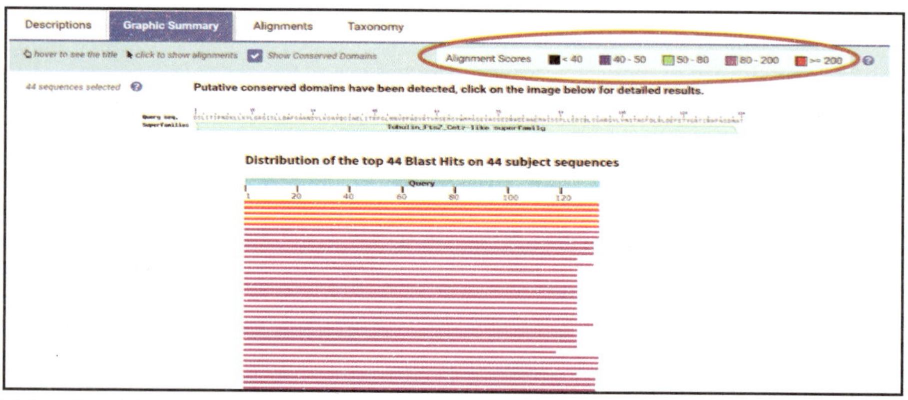

Figure 17: The Graphical summary for E. coli FtsZ protein alignment. The brown circle represents the color coding of the alignment score. The colored lines represent the different subject sequences, their length represents the query coverage and the color represents the alignment score. In this image, Subject 1 has a score of equal or greater than 200 and has 100% query coverage.

Upon clicking on the alignment tab, one can see the alignment between the query sequence and different subject sequence. One of the representative alignments is shown in figure 18. The example shows the alignment between E. coli FtsZ ad Bacillus subtilis FtsZ protein chain A. One can observe a line between the query and subject sequence with amino acid code and + sign which represents the identities and similarities among the sequences. The '+' sign represents the amino acids which similar physicochemical properties whereas the presence of a.a code represents the identities among two sequences. The mismatches between two sequences are represented as blank in the middle line. Further, for the given sequence out of 132 a.a, 69 a.a are identical. The positives represent the identical and similar a.a. There are 2 gaps in the given alignment and a score of 331. Thus, for all the alignment shown in the description, alignment files are generated.

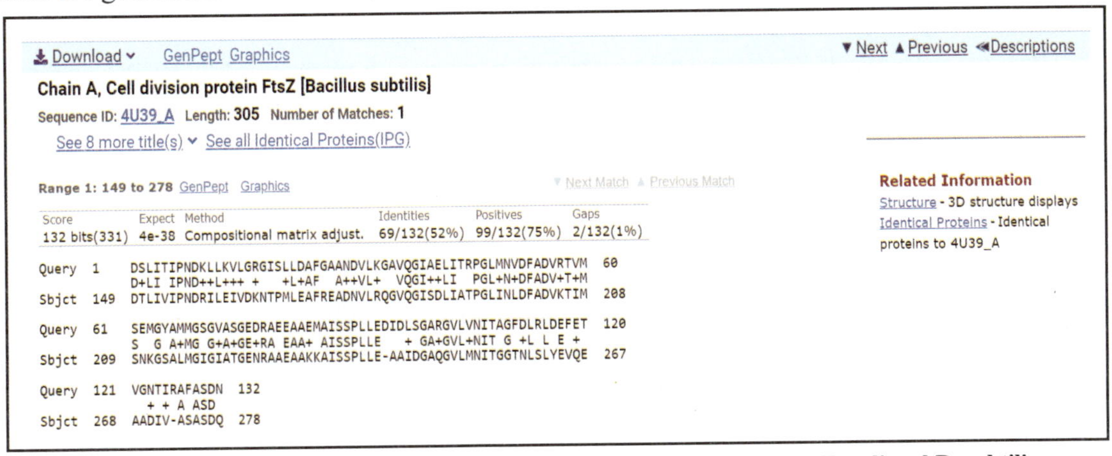

Figure 18: Alignment file observed after BLASTP for FtsZ protein from E. coli and B. subtilis.

Conclusion

In the current scenario with the focus on in silico work and use of computational tools for sequence analysis sequence alignment plays a very crucial role. Sequence alignment of the sequence results in analysis of structural, functional and evolutionary relationship between organisms. Nowadays, molecular modelling/ homology modelling is being used to construct the structure of proteins with the help of sequence itself. There has been development of various programs over the time for sequence alignment. In future, the importance of computational tools will gain more attention with modification in the existing programmes or algorithms and with the development of new algorithms.

Refrences

1. Cabanettes F, Klopp C. 2018. D-GENIES: dot plot large genomes in an interactive, efficient and simple way. PeerJ 6:e4958.
2. Altschul, S. F., Gish, W., Miller, W., Myers, E. W. and Lipman, D. J. 1990. Basic Local Alignment Search Tool. Journal of Molecular Biology 215: 403-410
3. Arslan AN, Eğecioğlu Ö. Dynamic programming-based approximation algorithms for sequence alignment with constraints. INFORMS Journal on Computing. 2004 Nov;16(4):441-58.

4. Boratyn GM, Camacho C, Cooper PS, Coulouris G, Fong A, Ma N, Madden TL, Matten WT, McGinnis SD, Merezhuk Y, Raytselis Y, Sayers EW, Tao T, Ye J, Zaretskaya I. BLAST: a more efficient report with usability improvements. Nucleic Acids Res. 2013 Jul;41.

5. Edgar RC, Batzoglou S. Multiple sequence alignment. Current opinion in structural biology. 2006 Jun 1;16(3):368-73.

6. Gertz EM, Yu YK, Agarwala R, Schäffer AA, Altschul SF. Composition-based statistics and translated nucleotide searches: improving the TBLASTN module of BLAST. BMC Biol. 2006 Dec 7;4:41. doi: 10.1186/1741-7007-4-41.

7. Gish, W., & States, D. J. Identification of protein coding regions by database similarity search. Nature Genetics 3, 266–272 (1993) doi:10.1038/ng0393-266.

8. Henikoff S, Henikoff JG. Amino acid substitution matrices from protein blocks. Proceedings of the National Academy of Sciences. 1992 Nov 15;89(22):10915-9.

9. Jia K, Jernigan RL. New amino acid substitution matrix brings sequence alignments into agreement with structure matches. Proteins: Structure, Function, and Bioinformatics. 2021 Jun;89(6):671-82.

10. Lobo, I. (2008) Basic Local Alignment Search Tool (BLAST). Nature Education 1(1):215.

11. Madeira F, Pearce M, Tivey AR, Basutkar P, Lee J, Edbali O, Madhusoodanan N, Kolesnikov A, Lopez R. Search and sequence analysis tools services from EMBL-EBI in 2022. Nucleic acids research. 2022 Jul 5;50(W1):W276-9.

4 Homology Modelling

Abstract

Homology modelling, also known as comparative modelling is based on the observation that protein structures are more conserved than protein sequences. If two proteins have similar sequences, they are likely to have similar structures. The basic steps of homology modelling include identifying a suitable template, aligning the target sequence with the template, building a model, and refining and validating the model. In this chapter we discussed various model building server, such as swiss model, Glaxy web, robetta. Structure validation programs such PROCHECK, ProSA and Verify3D is also discussed.

Introduction

All the living organisms are made up of biomolecules, such as carbohydrates, lipids, nucleic acids and proteins. Proteins are known to perform both structural and functional role. They support the structure of the cell and the enzymes work to support the function of the cell. Proteins are made up of amino acids. Amino acids are linked together through peptide bonds and the polymer is referred to as polypeptide.

Protein structural organization occurs at four different levels which includes primary structure, secondary structure, tertiary structure and quaternary structure (Figure 1). The protein primary structure includes the linear sequence of the amino acids. One end of the polypeptide chain contains the free amino group and another end contains the free carboxylic acid group. The primary structure is represented by the single letter code that represents the amino acids. The secondary structure is represented by alpha helices and beta sheets (parallel and anti-parallel) and represent the two-dimensional structure of the protein. The three-dimensional arrangement of the secondary structure in the space gives rise to the tertiary structure of the protein. The biological function of the protein is dictated by its tertiary structure. The oligomeric state (dimer, trimer or tetramer) formed by the aggregation of the compounds with tertiary structure result in quaternary structure.

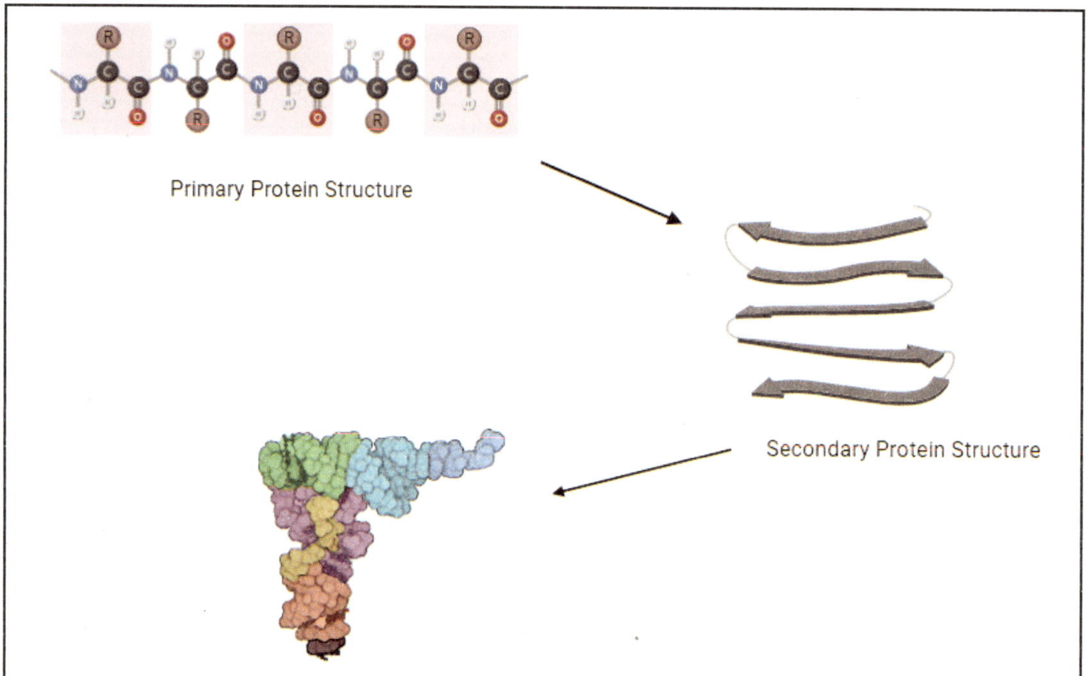

Figure 1: Pictorial representation of different types of protein structure.

The three-dimensional structure of the protein is very crucial for its function. There are very few techniques (Nuclear magnetic resonance and X-ray crystallography) to determine the structure of the proteins but are very expensive, complex and time consuming. Thus, use of computational tools to build the structural model of the protein is very useful, inexpensive and user friendly. There are mainly three types of in *silico*-based modelling methods for predicting the protein structure which includes ab initio, threading and homology modelling. Ab *initio* method predicts the protein structure using the amino acid sequence without any template. In case of threading, the three-dimensional structure of the protein is predicted by aligning the primary sequence of the protein to the primary sequence of the already available template in the protein data bank (PDB). Threading method is useful when no suitable template structures are available in the protein structure databases. Homology modelling is also referred to as comparative modelling and is based on the concept that the three-dimensional structure of the protein is more conserved than the primary structure of the protein.

Homology Modelling

Homology modelling as mentioned is dependent on the homology between two proteins. It is based on certain observations such as, proteins that are evolutionarily related have similar three-dimensional structure. Based on this, the three-dimensional protein model can be built from the similar proteins whose structure is known and which share a significant level of sequence similarity. Changes in the amino acid sequences does not change the structural domains of the proteins thereby not interfering with the protein function. It is assumed that proteins belonging to the same family will have the same structural domain which eventually results in homology modelling.

Homology modelling follows sequential steps which are repeated until and unless user gets the satisfactory model for the protein of interest (Figure 2). The steps include 1). Identification of homologous protein templates with known structure and folds based on sequence alignment. This step is based on sequence similarity between the target (protein of interest) and the template (with known structure); 2). Alignment of the protein sequences for both, the template and target protein; 3). Identification of structurally conserved regions and predicting the structure of the variable regions; 4) building of the three-dimensional model of the target protein; 5) Refinement and the optimization of the predicted model followed by assessment and validation.

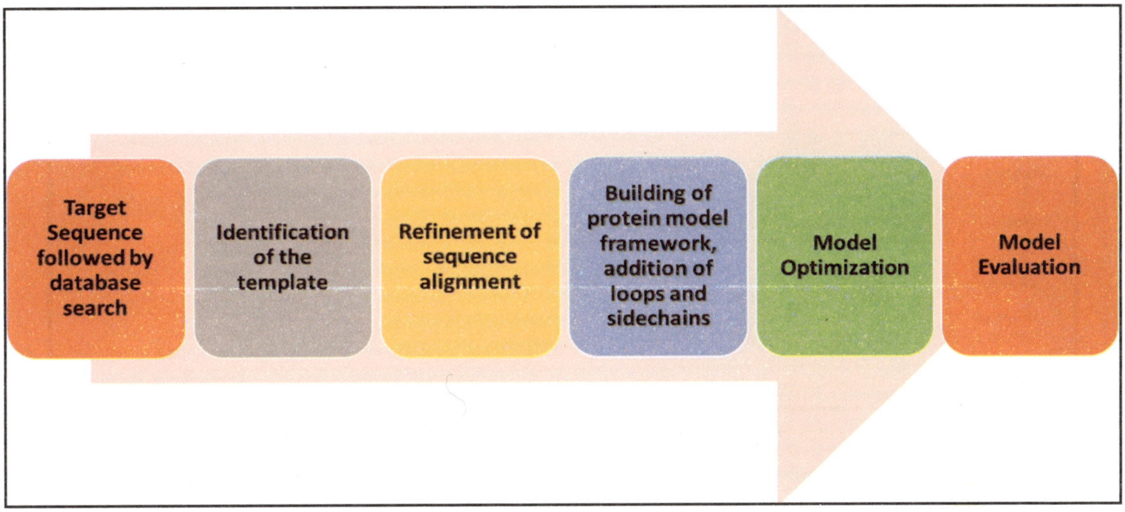

Figure 2: Represents the sequential steps that occurs during the homology modelling.

For target protein, having sequence homology to the template protein with solved three-dimensional structure, the model can be built based on the available structure. By rule, proteins with a minimum sequence identity of 25% is good for the construction of the protein models by homology. Sequence similarity above 40% and 50% provide good and excellent protein models, respectively (Maia et al., 2021).

For the sequence similarity, BLASTP is used and the search is done against PDB database. The higher the percentage of query coverage, i.e., the alignment between the target protein and the template, it is better for the homology modelling. Excellent coverage alignment is denoted by the query coverage of above 90% which always gives a higher score (Leach, 2001).

As discussed in sequence alignment chapter, gaps in the sequence alignment play an important role. The number and the length of the gaps is very crucial for the quality of the protein model. The higher the number of the gaps and the length, higher the chances of structural artifacts and less reliable is the modelled structure. Thus, while choosing the template, one should be careful about the gaps.

For example, consider the amylase protein from *Tetraodon nigroviridis* (Accession Id CAD20263.1). The sequence can be downloaded from ncbi (https://www.ncbi.nlm.nih.gov) or the accession id can be directly used for protein blast. While running the blast, one should use the protein data bank proteins as database and exclude the same organism for which alignment is being performed (Figure 3a). After running blastp we can observe the results with high homology to Japanese rice fish *(Oryzias latipes)* and humans *(Homo sapiens)* (Figure 3b). In the same table the accession number for template proteins from PDB can be found and used for homology modelling.

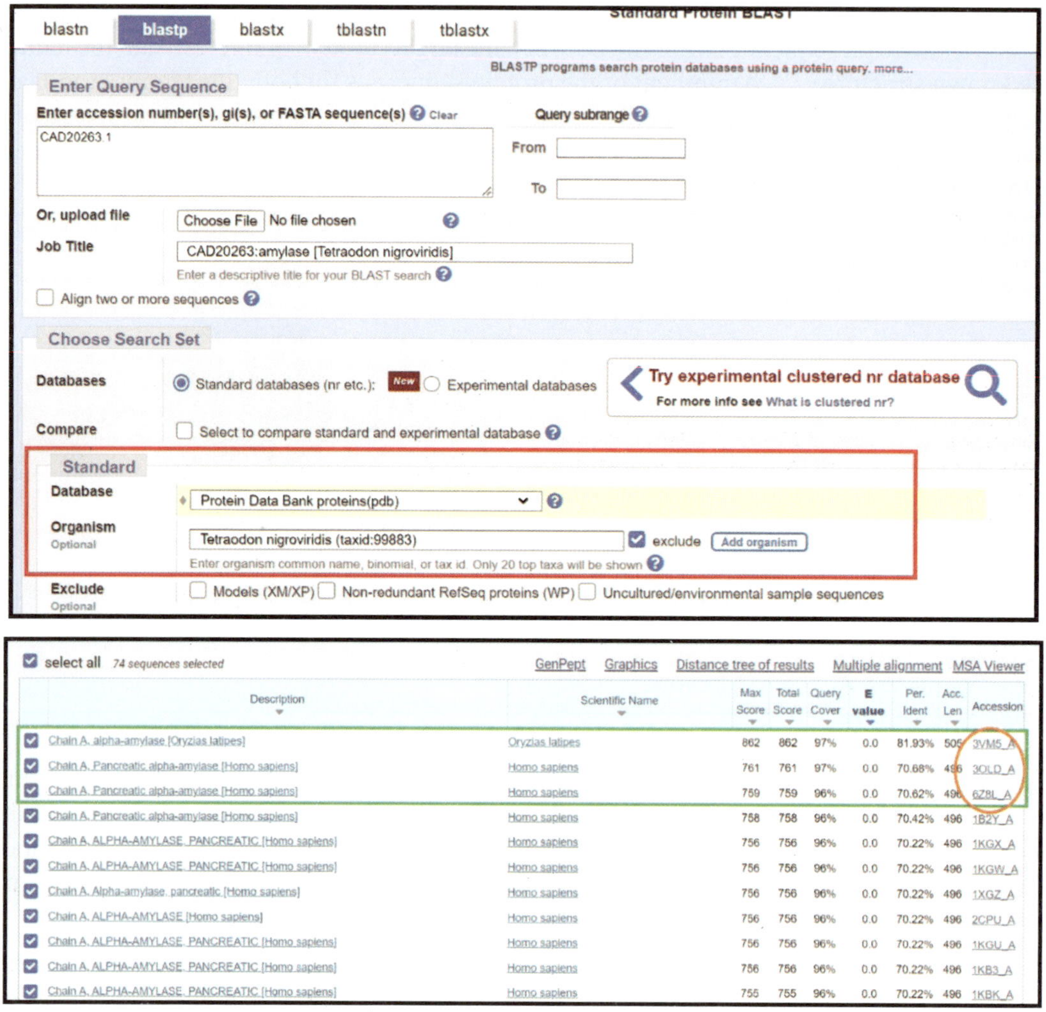

Figure 3: a) While performing BLASTP to see the alignment of the amino acid sequence of target protein, protein data bank proteins is selected in database and also the same organism must be excluded for alignment otherwise we will get 100% sequence similarity; b) Blast result indicating template identification based on sequence alignment indicating the query coverage, identity and PDB accession Id.

When we analyze the sequence similarity of the target protein and the template protein from *Oryzias latipes*, only 1 gap and 82% identity was observed (Figure 4). This indicates, when using this as a template we can get a reliable model based on homology modelling.

```
Chain A, alpha-amylase [Oryzias latipes]
Sequence ID: 3VM5_A  Length: 505  Number of Matches: 1

Range 1: 1 to 497  GenPept  Graphics          ▼ Next Match  ▲ Previous Match

Score           Expect  Method                          Identities     Positives      Gaps
862 bits(2227)  0.0     Compositional matrix adjust.    408/498(82%)   442/498(88%)   1/498(0%)

Query  16   QHTPHFKHGRTSIVHLFEWRWTDIAAECERFLGPNGFAGVQISPPNEHIILDSPWRPWWQ  75
            QH P+ + GRT+IVHLFEWRW DIAAECERFLGP GFAGVQISPPNEHI++ SPWRPWWQ
Sbjct  1    QHNPNTRDGRTAIVHLFEWRWADIAAECERFLGPKGFAGVQISPPNEHILVSSPWRPWWQ  60

Query  76   RYQPIGYNLCSRSGSENELRDMITRCNNVGVNIYVDVVINHMCRASSGEGTHSSCGSWFN  135
            RYQPI YNLCSRSG ENELRDMITRCNNVGVN+YVD VINHMC A  GEGTHSSCGSWFN
Sbjct  61   RYQPISYNLCSRSGGENELRDMITRCNNVGVNVYVDAVINHMCGAGGGEGTHSSCGSWFN  120

Query  136  AGTKDFPSVSYSYLDFNDHKCRTGSGNIESYDADVNQVRDCRLENLLDLALEKDYVRGKV  195
            A  KDFPSV YS LDFND KC+TGSGNIE+Y  D  QVRDCRL  LLDLALEKDYVRGKV
Sbjct  121  ANNKDFPSVPYSNLDFNDGKCKTGSGNIENY-GDPYQVRDCRLVGLLDLALEKDYVRGKV  179

Query  196  ADYMNRLVDMGVAGFRVDACKHMWPGDLSAVYGRLNNLNTNWFPGGSRPFIYQEVIDLGS  255
            AD+MN+L+DMGVAGFRVDACKHMWPGDL  VY RLNNLNT WFPGGSRPFI+QEVIDLG
Sbjct  180  ADFMNKLIDMGVAGFRVDACKHMWPGDLDNVYRRLNNLNTKWFPGGSRPFIFQEVIDLGG  239

Query  256  ELITAKEYFHLGRVTEFKYGAKLGTVFRKWNNEKLMYTKNWGEGWGFMPNGNAVVFIDNH  315
            E IT  EY  LGRVTEFKYGA+LG +FRKWN +KL YTKNWGEGWGFM +GNAVVF DNH
Sbjct  240  EPITTGEYVGLGRVTEFKYGARLGELFRKWNGQKLSYTKNWGEGWGFMADGNAVVFTDNH  299

Query  316  DNQRGHGAGGAAIVTFWDFRLHKMAVAYMLAHPYGVTRVMSSFRWNRHIVNGKDQNDWMG  375
            DNQRGHGAGGA+I+TFWD RL+KMAV YMLAHPYG TRVMSS+ W+R+ VNGKD+NDW+G
Sbjct  300  DNQRGHGAGGASILTFWDPRLYKMAVGYMLAHPYGFTRVMSSYSWDRNFVNGKDENDWIG  359

Query  376  PPSHPDGSTKSVPINPDETCGDGWVCEHRWRQIKNMVIFRNVVNGQPHSNWWDNNSNQVA  435
            PPS+ DGSTK VPINPD+TCGDGWVCEHRWRQI NMV FRNVVNGQPH+NWWDN +NQVA
Sbjct  360  PPSNGDGSTKPVPINPDQTCGDGWVCEHRWRQIMNMVQFRNVVNGQPHANWWDNGNNQVA  419

Query  436  FGRGNRGFIIFNNDDWDLDVTLNTGLPAGTYCDVISGQKEGGRCTGKQIHVGADGRAHFR  495
            FGRGNRGFI+FNNDDW LDVTLNTGLP GTYCDVISG K+GG CTGKQI VG DGRAHF
Sbjct  420  FGRGNRGFIVFNNDDWALDVTLNTGLPGGTYCDVISGNKDGGSCTGKQITVGGDGRAHFY  479

Query  496  ISNRDEDPFVAIHAESKL  513
            I+N +EDPF+AIHA+SKL
Sbjct  480  INNSEEDPFIAIHADSKL  497
```

Figure 4: Alignment of the target protein sequence and the template protein sequence identified through BlastP.

Once the template has been identified, the next step is the building of the three-dimensional protein model. For this, specific files are uploaded on specific programs or servers. Table 1 represents various tools for the generation of three-dimensional models.

Table 1: Free tools for the generation of three-dimensional protein model

S.No.	Name	Site	Type
1	Modeler	https://salilab.org/modeler/	Software
2	Swiss-model	https://swissmodel.expasy.org/	Server
3	Galaxy	http://galaxy.seoklab.org/	Server
4	ROBETTA	http://robetta.bakerlab.org/	Server
5	Phyre2	http://www.sbg.bio.ic.ac.uk/phyre2	Server
6	RaptorX	http://raptorx.uchicago.edu/	Server/ Software
7	CONFOLD	https://github.com/multicom-toolbox/CONFOLD	Software
8	I-TASSER	https://zhanggroup.org/I-TASSER/	Server

Swiss Model

The Swiss model server works with a minimum user input, for instance in the simplest scenario the user can provide the amino acid sequence of the target protein. There are other modes, which can be used for certain modelling cases where the user can select different template or can adjust the alignment between the target and the template (Waterhouse et al., 2018). There are three main interaction modes, first approach mode, alignment mode and project mode. In case of first approach mode, only the amino acid sequence of the protein is required. Suitable templates are automatically selected by the server. The modelling process initiates if at least one template has a sequence identity of more than 25% with the target sequence. The model reliability decreases with lower sequence identity.

The alignment mode provides more control over the modelling process by allowing the user to submit a predefined sequence alignment. The template and the target is specified by the user and the model is built based on the provided alignment.

The project mode is for advanced users. The input is in the form of DeepView project file which includes the superposed template structures and the alignment between the target and the template sequence. Various parameters such as template selection and gaps in the alignment. This model allows iterative refinement of the models, starting from the outputs of the 'first approach mode'.

Galaxy WEB

GalaxyWEB server runs on a cluster containing four linux servers which is equipped with 2.33 GHz Intel Xeon processors. Each processor consists of eight cores, providing a high level of parallel processing capability. The MySQL database is employed to manage and store data, ensuring efficient data handling and retrieval for the web application. For the prediction of structure, the proteins sequence needs to be submitted in FASTA format. For a 500-residue protein, the run time for the predicting the structure is approximately 7 hrs. After the run, five best models can be downloaded and viewed. All the models can be downloaded as a tar file (Ko et al., 2012).

For galaxy web, open the server using https://galaxy.seoklab.org. For protein structure prediction, three options are provided, template based analysis, loop prediction and domain prediction (Figure 7a). If you click on proteinTBM, new window opens. The job name can be provided along with mail id as the run time is approximately 7 hrs and the result link can directly be mailed. The amino acid sequence for Galaxy web is between 30-1000 amino acids. The user can either paste the sequence directly in the space provided or can upload the sequence file in the given format (Figure 7b).

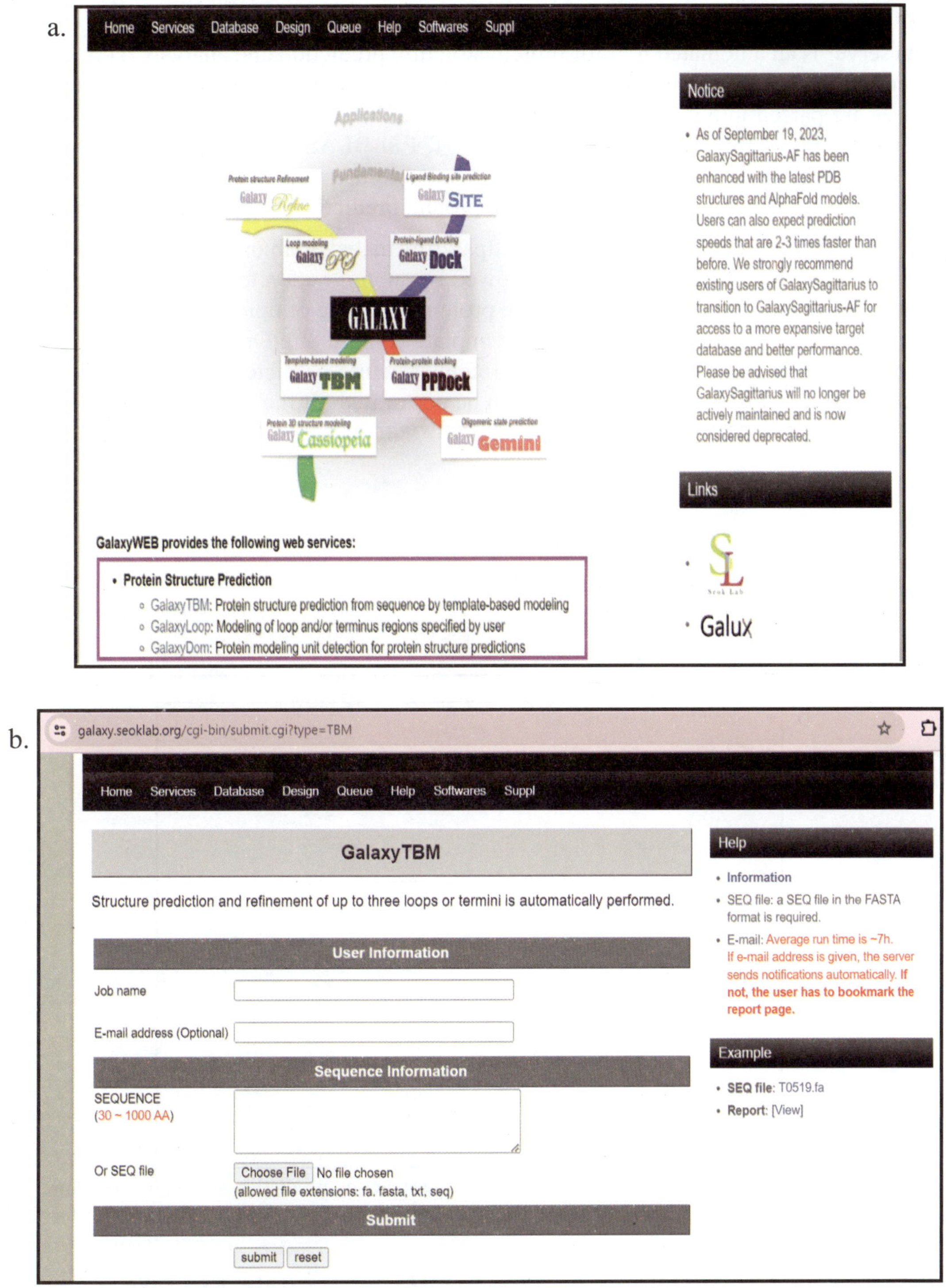

Figure 6. a) Represents the homepage of galaxy web b) represent the run page for template-based model preparation.

Robetta

Robetta server is an automated tool for protein structure prediction and analysis. This server is designed to predict the protein structure with minimal user intervention. In this, the submitted sequences are parsed into putative domains, which are the segments of proteins that can fold independently into stable structures. It uses BLAST, PSI-BLAST, FFAS03 or 3D-Jury to find matches for proteins of known structures. If no suitable template is found, Robetta uses the Rosetta de novo fragment insertion method for protein prediction (Kim et al., 2004).

As we open the homepage of Robetta (https://robetta.bakerlab.org), it gives an option of protein structure prediction click on that followed by submit button to start a new job (Figure 8a). On the next page, we can either paste the protein sequence or upload the same in the fasta format. We can also specify the PDB file as template (Figure 8b). Once we submit the job, it will take sometime to give the results. The protein structure predicted can be downloaded in PDB format and validation and refinement can be performed for the same.

a.

b.

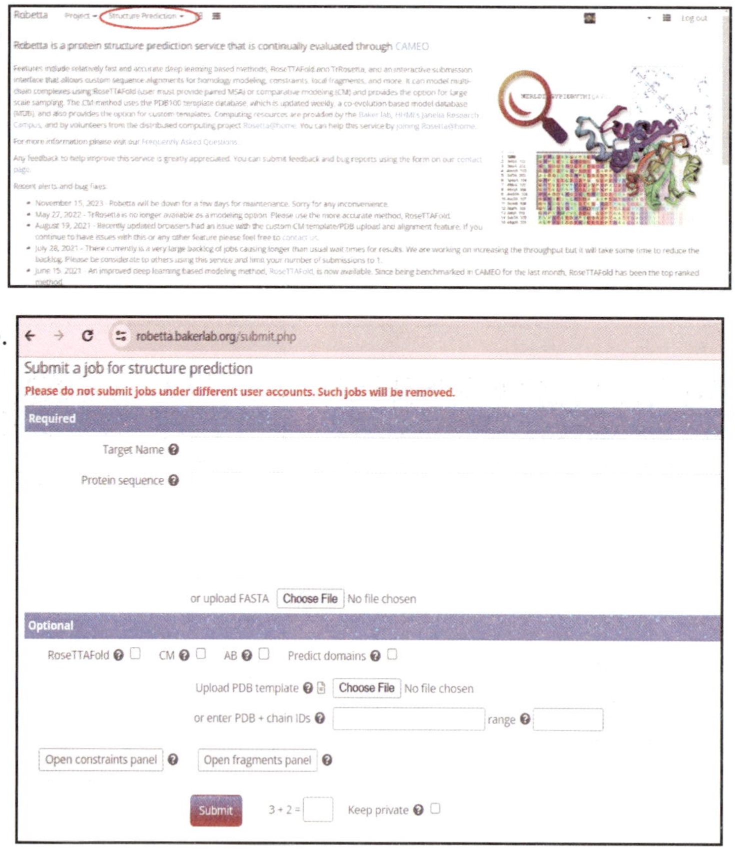

Figure 7: a) Robetta home page, the red circle represents the structure prediction option. b) Job submission page of Robetta where one can either paste the sequence or upload it and provide the PDB template for model generation.

Validation and Refinement of The Predicted Protein Structure

Homology models, being theoretical-computational approximations of real protein structures, need thorough validation and often refinement to ensure their accuracy and reliability. One of the most widely used tools for this purpose is the Ramachandran plot, which helps assess the stereochemical quality of protein structures.

The Ramachandran plot evaluates the phi (φ) and psi (ψ) dihedral angles of the amino acid residues in a protein structure. It identifies allowed and disallowed regions for these angles, based on known protein structures. Ideally, most residues should fall within the allowed regions, indicating a stereochemically plausible structure.

Here we checked the Ramachandran plot for the model 1 generated in figure 6. Open https://saves.mbi.ucla.edu/, it has 5 programmes to validate the protein structure and one of them is PROCHECK which helps in the analysis of Ramachandran plot (Laskowski et al., 1993). Upload the PDB file downloaded after model building and run programs. Click on start for the programs which you want to run and wait for few minutes until the program finishes after that click on results. The results of Ramachandran Plot is shown in figure 9. Around 90% of the amino acids are present in the most favoured region. One can further go for refinement if required.

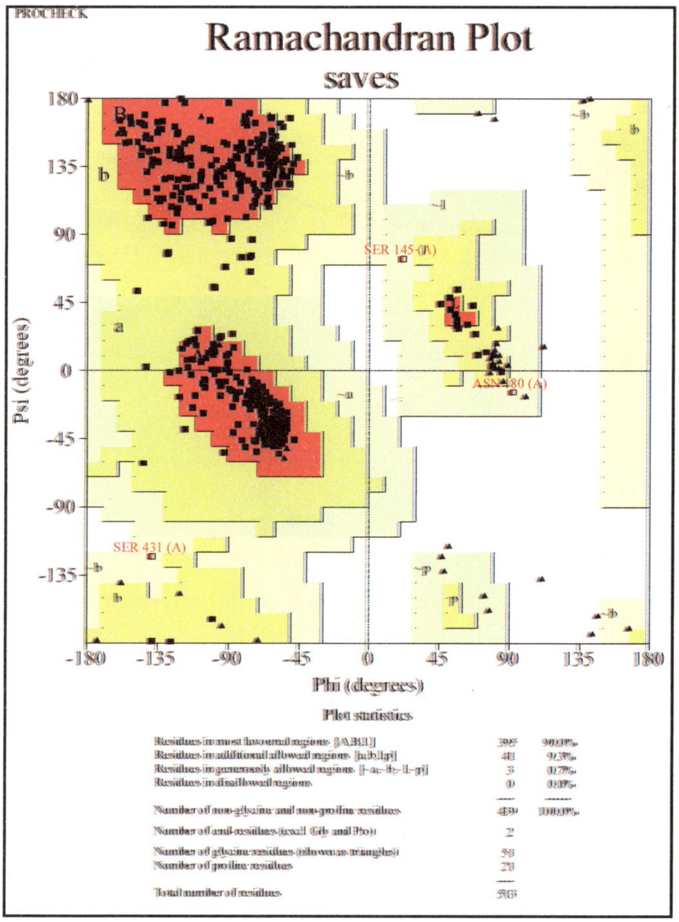

Figure 8: Ramachandran plot where the 90% of the amino acids are in most favoured region represented as black dots in red region, 9.3% in additional allowed region and 0.7% in generously allowed region. A good model should have more than 90% of the amino acids in the most favoured region.

ProSA (Protein Structure Analysis) web is a tool used for evaluating the quality of protein structures. It provides a comprehensive analysis of the protein model by comparing its quality scores against known protein structures (Wiederstein et al., 2007). The quality score is depicted by Z-score and Energy Plot. The Z-score indicates the overall quality of a protein structure. It measures the deviation of the total energy of the structure from the mean energy of a set of high-resolution structures. A Z-score outside the range characteristic of native proteins suggests a problematic structure. ProSA web calculates the energy for each residue and plots these values along the sequence referred to as energy plot. In the energy plot negative values generally indicate stable, well-folded regions, while positive values highlight potentially problematic areas.

For ProSA, open the server using https://prosa.services.came.sbg.ac.at/prosa.php. At the home page upload the PDB file of the model and click on analyse. The results are presented in graphical format for Z score and for the energy plot (Figure 10).

a.

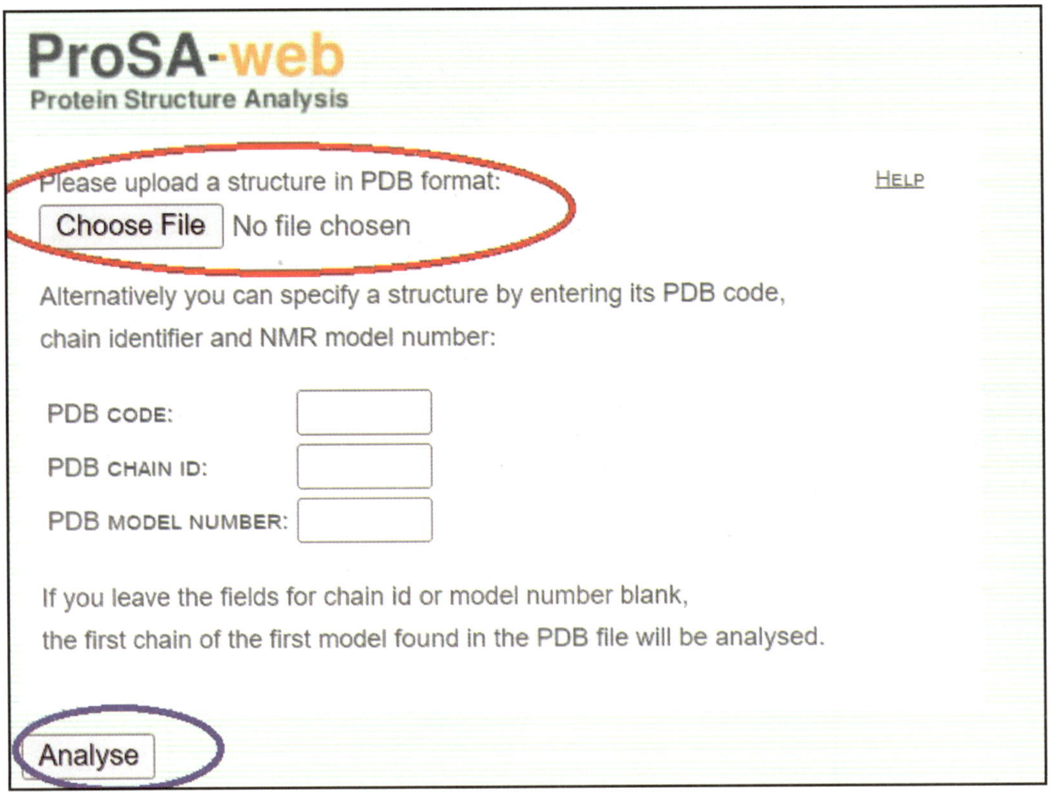

b.

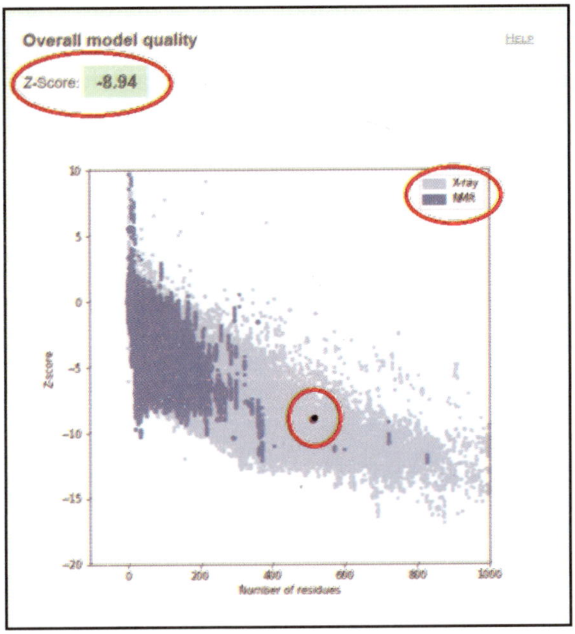

c.

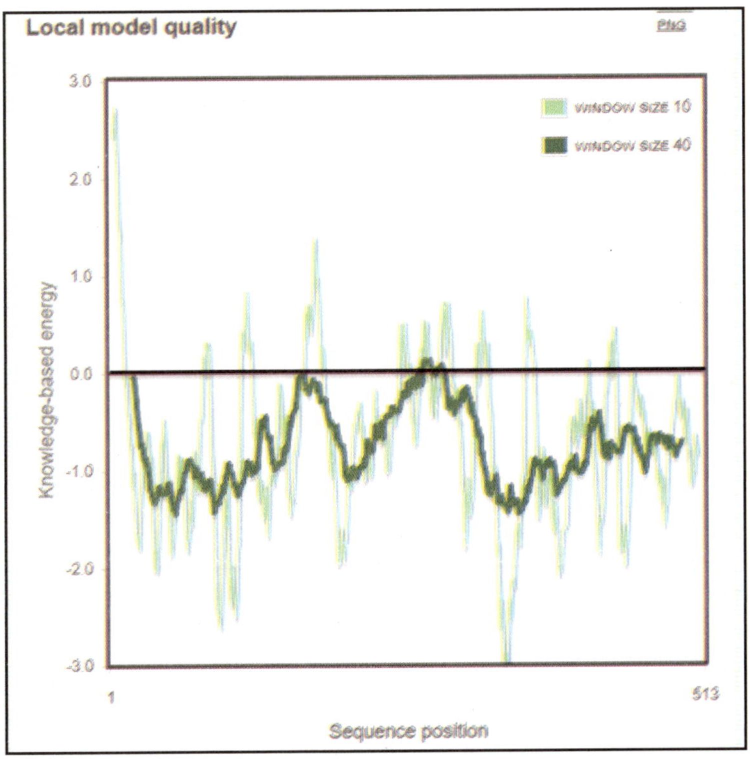

Figure 9: a) Represents the homepage of ProSA web where one can upload the pdb file of the protein structure. b) Represents the Z score of the protein to be -8.94. The Z score falls within the range characteristics of native proteins indicating a good structure. c) Represents the energy plot. Most of the amino acids fall in the negative region indicating stable, well-folded regions. A very small region is in positive range and thus, if required we can go for minor refinement.

The VERIFY3D server is a powerful tool for local quality analysis of protein structures. It assesses the compatibility of an atomic model (3D) with its own amino acid sequence (1D) to ensure that the local environments of residues are realistic (Lüthy et al., 1992; Wallner and Elofsson, 2006). This process helps identify specific regions of low quality in a protein model, which can then be targeted for further refinement. The verify 3D scores reflect how well the 3D structure of each residue matches its preferred environment in high-quality structures. Residues scoring above a threshold (usually around 0.2) are considered compatible with their environment, while those scoring below this threshold indicate potential issues.

For running VERIFY3D, open SAVES (https://saves.mbi.ucla.edu/). Upload the protein pdb file and run the program. If around 80% of the amino acids have a 3D/1D>=0.1 are considered pass (Figure 11).

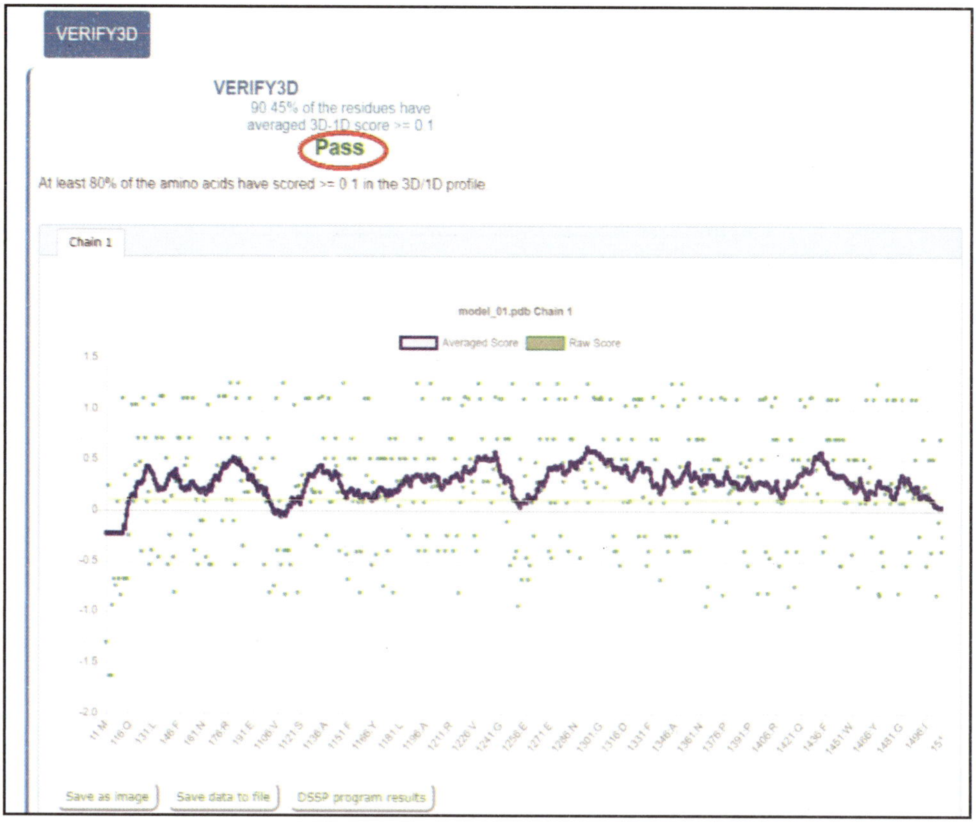

Figure 10: Represents VERIFY3D graph for the protein model. The result indicates 'pass' highlighted by red circle.

Galaxy Refine

GalaxyRefine is a web server used for the refinement of protein structures, particularly homology models. It enhances the quality of initial protein models by optimizing both side-chain conformations and the overall protein backbone (Heo et al., 2016). This process helps to correct errors and improve the stereochemical and energetic properties of the protein model. The initial step involves perturbing the protein structure to sample alternative conformations. This includes small displacements of side chains and the backbone to explore the conformational space. Using a side-chain repacking algorithm, GalaxyRefine optimizes the positions of side chains to reduce steric clashes and improve the overall fit of the side chains in their local environments. The backbone of the protein is then relaxed through restrained energy minimization. This step helps in relieving any strain and bringing the structure closer to its native state. The process involves multiple iterations of side-chain repacking and backbone relaxation to progressively improve the quality of the protein model. A final round of energy minimization is conducted to ensure that the refined structure has the lowest possible energy, indicating a stable conformation. The refined models are evaluated based on various quality metrics, including the GDT-HA (Global Distance Test High Accuracy) score, RMSD (Root Mean Square Deviation), and MolProbity scores, which assess the stereochemical quality and atomic interactions within the protein.

One can open galaxy refine using the link https://galaxy.seoklab.org/cgi-bin/submit.cgi?type=REFINE. PDB file whose structure needs to be refined needs to be uploaded and one should click on submit. The total run takes approximately 1-2 hrs. The run page is represented in figure 12a. The result page is shown in figure 12b with the scores, RMSD and MolProbity score. After refinement one can further validate the structure using PROCHECK, VERIFY3D and ProSA.

a.

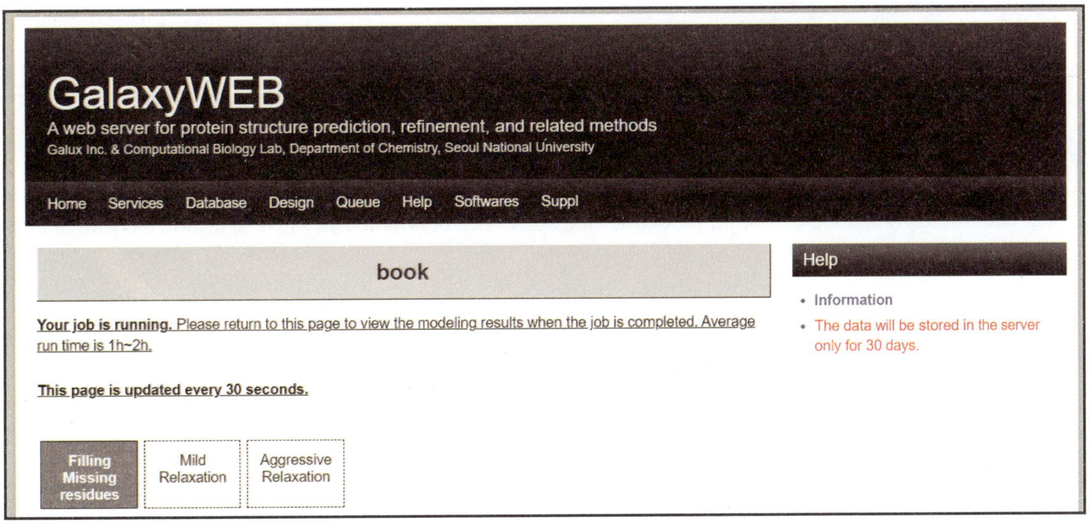

b.

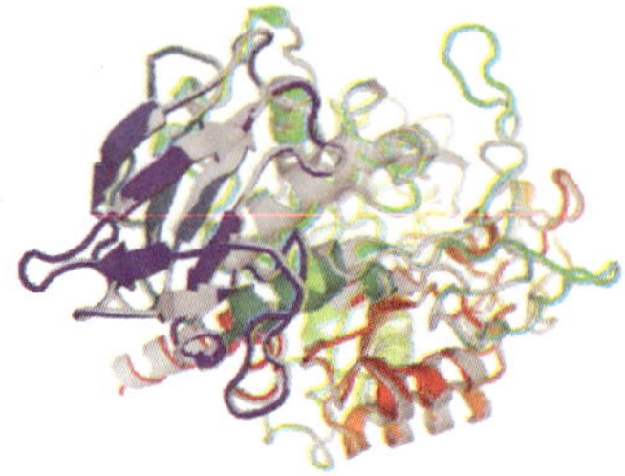

View in PV [Model 1] [Model 2] [Model 3] [Model 4] [Model 5] [All Models]

Download [Model 1] [Model 2] [Model 3] [Model 4] [Model 5] [All Models]

Structure Information

Model	GDT-HA	RMSD	MolProbity	Clash score	Poor rotamers	Rama favored
Initial	1.0000	0.000	1.294	3.1	0.2	98.9
MODEL 1	0.9942	0.242	1.601	12.3	0.5	99.0
MODEL 2	0.9956	0.234	1.649	13.9	0.5	98.6
MODEL 3	0.9917	0.246	1.663	14.3	0.7	98.8
MODEL 4	0.9932	0.232	1.653	14.0	0.0	98.8
MODEL 5	0.9937	0.231	1.642	13.6	0.5	99.0

Download

- 5 refined models [DOWNLOAD]

Figure 11: a) Represents the progress of the refinement process after submission of the job. The run takes a total of 1-2 hours. b) Represents the result page. After analysing the score, RMSD value and MolProbity score select the model and further validation can be performed for the same. The best 5 models can also be downloaded in pdb file format.

Application of Homology Modelling

Homology modelling is a versatile and powerful tool that extends our ability to understand protein structures and functions in the absence of experimental data. Its applications span drug discovery, functional annotation, protein engineering, and beyond, making it an indispensable technique in modern molecular biology and bioinformatics.

In context with Drug discovery and design homology models help in identification and validation of the potential drug targets by providing structural information about proteins involved in disease pathways. These models can be used in virtual screening to identify potential drug candidates from large compound libraries (Evers et al., 2003). Beside screening, the binding sites of the potential drugs can be predicted which facilitates the designing and optimization of small molecules or peptides (Hillisch et al., 2004).

During functional annotation of proteins, by modelling the active sites of the enzymes, researchers can predict the substrate specificity and catalytic mechanisms. Structural models also help to understand the protein-protein interaction which is crucial for elucidating cellular processes and signalling pathways.

Homology models can reveal how mutations affect protein structure and function, providing insights into the protein function and molecular basis of diseases (Hamano et al., 2020). The structural models can help to predict the effects of genetic variation in patients, aiding in the development of the personalized treatment strategies. Homology modelling can also be used in protein engineering where proteins with improved or novel functions can be designed by identifying key residues for mutagenesis. Further mutations can be designed to enhance the thermal or pH stability of proteins which is important for industrial and therapeutic applications.

Models help in tracing the evolutionary history of proteins by comparing the structures of the homologous proteins from different proteins (Suresh and Das 2014). Homology modelling guides the engineering of enzymes for various industrial applications, such as biofuel production and waste management.

Conclusion

Homology modelling is particularly useful when experimental methods such as X-ray crystallography or nuclear magnetic resonance are not feasible or have not been successful in determining the structure of the target protein. It has become an essential tool in structural biology and drug discovery, allowing researchers to generate structural hypotheses and gain insights into the molecular mechanisms of proteins. However, it's important to note that the accuracy of homology models depends on the quality of the sequence alignment and the similarity between the target and template proteins. Additionally, homology modelling may not accurately capture structural features such as flexible loops or domain movements that are not conserved between the target and template proteins.

References

1. Evers A, Gohlke H, Klebe G. Ligand-supported homology modelling of protein binding-sites using knowledge-based potentials. Journal of molecular biology. 2003 Nov 21;334(2):327-45.
2. Hamano F, Matoba K, Hashidate-Yoshida T, Suzuki T, Miura K, Hishikawa D, Harayama T, Yuki K, Kita Y, Noda NN, Shimizu T. Mutagenesis and homology modeling reveal a predicted pocket of lysophosphatidylcholine acyltransferase 2 to catch Acyl-CoA. bioRxiv. 2020 Nov 2:2020-10.
3. Heo L, Lee H, Seok C. GalaxyRefineComplex: Refinement of protein-protein complex model structures driven by interface repacking. Scientific reports. 2016 Aug 18;6(1):32153.
4. Hillisch A, Pineda LF, Hilgenfeld R. Utility of homology models in the drug discovery process. Drug discovery today. 2004 Aug 1;9(15):659-69.
5. Kim DE, Chivian D, Baker D. Protein structure prediction and analysis using the Robetta server. Nucleic acids research. 2004 Jul 1;32(suppl_2):W526-31.
6. Ko J, Park H, Heo L, Seok C. GalaxyWEB server for protein structure prediction and refinement. Nucleic acids research. 2012 May 30;40(W1):W294-7.
7. Laskowski RA, MacArthur MW, Moss DS, Thornton JM. PROCHECK: a program to check the stereochemical quality of protein structures. Journal of applied crystallography. 1993 Apr 1;26(2):283-91.

8. Leach AR. Molecular modelling: principles and applications. Pearson education; 2001.
9. Lüthy R, Bowie JU, Eisenberg D. Assessment of protein models with three-dimensional profiles. Nature. 1992 Mar 5;356(6364):83-5.
10. Maia RT, de Moraes Filho RM, Campos MD, editors. Homology Molecular Modeling: Perspectives and Applications. BoD–Books on Demand; 2021 Mar 10.
11. Suresh N, Das A. Homology modelling and phylogenetic analysis of alkaline phosphatase, acid phosphatase and phytase genes from Aspergillus fumigatus. Rasayan Journal of Chemistry. 2014;7:201-6.
12. Wallner B, Elofsson A. Identification of correct regions in protein models using structural, alignment, and consensus information. Protein Science. 2006 Apr;15(4):900-13.
13. Waterhouse A, Bertoni M, Bienert S, Studer G, Tauriello G, Gumienny R, Heer FT, de Beer TA, Rempfer C, Bordoli L, Lepore R. SWISS-MODEL: homology modelling of protein structures and complexes. Nucleic acids research. 2018 Jul 2;46(W1):W296-303.
14. Wiederstein M, Sippl MJ. ProSA-web: interactive web service for the recognition of errors in three-dimensional structures of proteins. Nucleic acids research. 2007 Jul 1;35(suppl_2):W407-10.

5 Computational Biology

Computational biology is an interdisciplinary field that applies algorithms, data analysis, mathematical modeling, and simulations to understand and model the structures and processes of biological systems. This field encompasses various types of biological data, including genomic, proteomic, and metabolomic data, and aims to provide insights into complex biological phenomena and solve biological problems through computational methods.

Historical Background and Evolution

The origins of computational biology can be traced back to the early 20th century, with roots in the mathematical modeling of biological processes. Early pioneers like Alan Turing and Norbert Wiener laid the groundwork for using mathematical and computational techniques to understand biological systems. Turing's work on morphogenesis and Wiener's contributions to cybernetics highlighted the potential of applying computational methods to biological questions.In the 1950s and 1960s, the advent of computers provided a new dimension to biological research. One of the first significant milestones was the development of sequence alignment algorithms. Margaret Dayhoff and her colleagues created the first protein sequence database and introduced the concept of the PAM (Point Accepted Mutation) matrix, which was a pioneering effort in sequence comparison.

During the 1960s, the field of bioinformatics began to take shape. The development of the Needleman-Wunsch algorithm for optimal sequence alignment in 1970 marked a significant advancement. This algorithm laid the foundation for many subsequent developments in sequence analysis. Parallel to these efforts, the first attempts to predict protein secondary structure using computational methods were made, highlighting the potential for computational tools in structural biology.

The 1980s and 1990s saw an explosion of computational tools and techniques, driven by the increasing availability of biological data. The Human Genome Project, launched in 1990, was a major catalyst for the field. This international effort aimed to sequence the entire human genome, and it spurred the development of new algorithms, data structures, and databases to handle the vast amounts of genetic data being generated. Key developments during this period included the creation of GenBank, a comprehensive repository for nucleotide sequences, and the Protein Data Bank (PDB), which stores three-dimensional structures of proteins and nucleic acids. Advances in algorithmic techniques, such as the BLAST (Basic Local Alignment Search Tool) algorithm developed by Altschul et al. in 1990, revolutionized the way scientists searched for sequence similarities.

The turn of the century brought further integration of computational methods with biological research. The completion of the Human Genome Project in 2003 marked the beginning of a new era in genomics. High-throughput sequencing technologies, often referred to as next-generation sequencing (NGS), allowed for rapid and cost-effective sequencing of entire genomes, producing data at an unprecedented scale. During this period, computational biology expanded to include systems biology, which seeks to understand the complex interactions within biological systems. The development of computational models to simulate these interactions became a cornerstone of the field. Researchers began to construct network models of cellular processes, integrating data from genomics, proteomics, and metabolomics to create comprehensive models of biological systems. Machine learning and artificial intelligence (AI) have also become integral to computational biology. Techniques such as deep learning are now used to analyze and interpret complex biological data, from predicting protein structures to identifying genetic variants associated with diseases. The rise of personalized medicine, which tailors' medical treatment to the individual characteristics of each patient, relies heavily on computational tools to analyze genetic data and predict responses to treatment.

Important and Application in Modern Biology

Computational biology plays a crucial role in modern biology by enabling researchers to analyze large datasets, model complex biological systems, and predict the behaviour of these systems under various conditions. Applications include understanding genetic diseases, drug discovery, personalized medicine, and the study of evolutionary processes. By integrating diverse data types and applying sophisticated computational techniques, computational biology has become indispensable in advancing biological research and healthcare. Today, computational biology continues to evolve rapidly, driven by advancements in technology and an ever-growing body of biological data. The integration of multi-omics data, which combines genomic, transcriptomic, proteomic, and metabolomic information, is a major focus. This holistic approach aims to provide a more complete understanding of biological systems and their regulation. Emerging technologies such as CRISPR-Cas9 for genome editing and single-cell sequencing are generating new types of data, further expanding the scope of computational biology. Researchers are also exploring the use of quantum computing to tackle complex biological problems that are currently computationally intractable. The future of computational biology lies in its ability to integrate and analyze diverse data types, develop more accurate predictive models, and translate these findings into practical applications in medicine, agriculture, and environmental science. As computational power continues to increase and algorithms become more sophisticated, the potential for computational biology to revolutionize our understanding of life and improve human health is immense.

Core Concept And Technique

Algorithms and Data Structures

Algorithms and data structures form the backbone of computational biology, enabling efficient processing and analysis of biological data. Key algorithms include those for sequence alignment, phylogenetic analysis, and structural prediction. Data structures such as trees, graphs, and hash tables are used to organize and retrieve biological information quickly and accurately.

Sequence Alignment Algorithms

Sequence alignment is fundamental in computational biology for comparing DNA, RNA, or protein sequences. The goal is to identify regions of similarity that may indicate functional, structural, or evolutionary relationships.

Global and Local Alignment

- **Needleman-Wunsch Algorithm:** This dynamic programming algorithm performs global alignment, which aligns sequences along their entire length. It is particularly useful when the sequences are of similar length and highly homologous. The algorithm constructs a matrix and traces back to find the optimal alignment, ensuring the highest possible score based on a given scoring system.
- **Smith-Waterman Algorithm:** Unlike global alignment, the Smith-Waterman algorithm performs local alignment, which finds the best matching subsequence within the sequences. This is useful for aligning sequences with varying lengths or identifying conserved regions within long sequences. The algorithm also uses dynamic programming to build a matrix and traceback to identify the highest-scoring local alignments.

Heuristic Methods

- **BLAST (Basic Local Alignment Search Tool):** BLAST is a widely used heuristic algorithm for comparing an input sequence (query) against a database of sequences. It balances speed and accuracy by finding short, high-scoring sequence matches (seeds) and extending them. This method is less computationally intensive than full dynamic programming approaches and is suitable for large-scale database searches.

Phylogenetic Analysis

Phylogenetic analysis involves reconstructing the evolutionary relationships between different species or genes. This is often represented in the form of a phylogenetic tree.

- **Neighbour-Joining Algorithm:** This is a distance-based method for constructing phylogenetic trees. It iteratively joins the closest pair of nodes (species or sequences) and recalculates distances to form a new node until a single tree is constructed. It is efficient and suitable for large datasets.
- **Maximum Likelihood and Bayesian Methods:** These probabilistic methods provide more accurate phylogenetic trees by considering different possible trees and selecting the one with the highest likelihood given the data. They use sophisticated statistical models to account for various evolutionary factors. Tools like PhyML and MrBayes implement these methods.

Protein Structure Prediction

Predicting the three-dimensional structure of proteins from their amino acid sequences is a major challenge in computational biology.

- **Homology Modeling:** This method predicts a protein's structure based on the known structures of homologous proteins. By aligning the target sequence with the template structures, researchers can infer the 3D structure of the protein of interest. Tools like MODELLER are commonly used for homology modeling.

- **Threading:** Threading or fold recognition attempts to fit the target sequence directly onto a structural template from a database of known folds. It is useful when homologous structures are not available. Programs like Phyre2 and I-TASSER utilize threading techniques.
- **Ab Initio Methods:** When no suitable templates are available, ab initio methods predict protein structure from scratch, based solely on physical and chemical principles. These methods, such as those used in the Rosetta software, rely heavily on computational power and sophisticated algorithms to explore possible conformations.

Data Structures

Efficient data structures are essential for storing, retrieving, and managing biological data.

Tree and Graph

- **Suffix Trees and Arrays:** These data structures are used for efficient string matching and sequence alignment. A suffix tree is a compressed trie of all suffixes of a given text, allowing rapid searches for substring matches. Suffix arrays are a space-efficient alternative that provides similar functionality.
- **Phylogenetic Trees:** Representing evolutionary relationships, phylogenetic trees use nodes and edges to model the divergence of species or genes from common ancestors. Different algorithms generate these trees, which can be visualized and analyzed using tools like MEGA or FigTree.
- **Hash Tables:** These are used for fast data retrieval, such as in the k-mer counting process in genomics. A hash table maps keys (e.g., short DNA sequences) to values (e.g., their frequencies), enabling rapid lookups and updates.

Graphs

- **Graph Databases:** Graphs are used to represent complex biological networks, such as protein-protein interaction networks, metabolic pathways, and gene regulatory networks. Nodes represent entities (proteins, metabolites, genes), and edges represent interactions. Tools like Cytoscape are used to analyze and visualize these biological networks.

The algorithms and data structures described are foundational for many applications in computational biology:

- **Genome Assembly:** Algorithms like de Bruijn graphs are used to assemble short sequencing reads into complete genomes, handling the complexity of overlapping reads and repeats.
- **Gene Prediction:** Hidden Markov Models (HMMs) are employed to predict gene locations in genomic sequences by modeling the statistical properties of coding and non-coding regions.
- **Drug Discovery:** Computational docking algorithms predict how small molecules interact with target proteins, aiding in the design of new drugs.

Algorithms and data structures are essential tools in computational biology, enabling researchers to analyze vast amounts of biological data efficiently. By leveraging these computational techniques, scientists can uncover insights into biological processes, evolutionary relationships, and molecular structures, driving advancements in genomics, systems biology, and personalized medicine. As the field continues to evolve, the development of more sophisticated algorithms and data structures will be crucial for addressing increasingly complex biological questions.

Machine Learning and Artificial Intelligence in Biology

Machine learning and artificial intelligence (AI) have revolutionized computational biology by providing tools to uncover patterns and make predictions from complex biological data. Techniques such as neural networks, support vector machines, and clustering algorithms are employed to classify data, predict protein structures, and analyze gene expression profiles. AI-driven approaches are particularly valuable in personalized medicine, where they help tailor treatments to individual genetic profiles.Machine learning involves the development of algorithms that can learn from and make predictions or decisions based on data. AI encompasses a broader range of technologies, including ML, that enable machines to perform tasks that typically require human intelligence.

Figure 1: workflow depicting the steps involved in a computational biology analysis pipeline

Types of Machine Learning

- **Supervised Learning:** In supervised learning, algorithms are trained on labeled data, where the input-output pairs are known. The goal is to learn a mapping from inputs to outputs to make accurate predictions on new, unseen data. Common algorithms include linear regression, support vector machines (SVM), and neural networks.
- **Unsupervised Learning:** Unsupervised learning algorithms identify patterns or structures in data without labeled outputs. Techniques such as clustering (e.g., k-means, hierarchical clustering) and dimensionality reduction (e.g., principal component analysis, t-SNE) are used to uncover underlying data structures.
- **Reinforcement Learning:** In reinforcement learning, algorithms learn by interacting with an environment to maximize a reward signal. This approach is used in applications where decisions need to be made sequentially, such as in robotic control or game playing.

Deep Learning

Deep learning, a subset of ML, involves neural networks with many layers (deep neural networks). These models are particularly effective at handling large and complex datasets, making them well-suited for tasks like image recognition, natural language processing, and genomic data analysis.

Machine learning and AI have numerous applications across different areas of biology, enhancing our ability to understand and manipulate biological systems.

- **Gene Expression Analysis:** ML algorithms are used to analyze gene expression data from technologies like RNA-seq. Supervised learning methods can identify genes that are differentially expressed between conditions, while unsupervised methods can cluster genes with similar expression patterns.

- **Variant Calling and Genome Annotation:** Deep learning models, such as convolutional neural networks (CNNs), are employed to identify genetic variants from sequencing data and predict the functional effects of these variants. Tools like DeepVariant have shown high accuracy in variant calling.
- **Protein Structure Prediction:**AlphaFold, a deep learning-based system developed by DeepMind, has made significant strides in predicting protein structures from amino acid sequences. It uses neural networks trained on known protein structures to predict the 3D configurations of new proteins.
- **Protein Function Prediction:** ML algorithms can predict the functions of uncharacterized proteins by analyzing sequence data, structural information, and interaction networks. Methods such as SVMs and random forests are commonly used for these tasks.
- **Virtual Screening:** AI models can screen large libraries of compounds to identify potential drug candidates. By predicting the binding affinity between small molecules and target proteins, these models help prioritize compounds for experimental testing.
- **Drug Repurposing:** ML techniques analyze existing drug databases to find new therapeutic uses for approved drugs. By identifying patterns and similarities in drug-target interactions, AI can suggest alternative treatments for diseases.
- **De Novo Drug Design:** Generative models, including variationalautoencoders (VAEs) and generative adversarial networks (GANs), are used to design novel drug molecules with desired properties. These models generate new chemical structures that can be synthesized and tested for efficacy.
- **Network Inference:** ML algorithms reconstruct biological networks, such as gene regulatory networks and protein-protein interaction networks, from high-throughput data. Techniques like Bayesian networks and network-based inference methods are used to predict interactions and regulatory relationships.
- **Pathway Analysis:** AI tools analyze metabolic and signaling pathways to understand their regulation and dynamics. These analyses help identify key regulators and potential therapeutic targets.
- Predictive Modeling: ML models predict disease risk, treatment response, and patient outcomes based on genomic, clinical, and lifestyle data. These predictions enable personalized treatment plans tailored to individual patients.
- **Electronic Health Records (EHR) Analysis:** AI algorithms analyze EHR data to identify patterns associated with diseases, predict disease progression, and recommend personalized interventions. Natural language processing (NLP) techniques extract relevant information from unstructured clinical notes.

Challenges and Future Direction

While ML and AI have transformative potential in biology, several challenges need to be addressed to realize their full potential:

- **Data Quality and Quantity:** High-quality, well-annotated datasets are essential for training accurate models. However, biological data can be noisy, incomplete, and heterogeneous, posing challenges for model development.
- **Interpretability:** Many ML models, especially deep learning models, are often considered "black boxes" due to their complexity. Developing interpretable models that provide insights into their predictions is crucial for gaining trust and understanding in biological research.
- **Generalizability:** Ensuring that models generalize well to new, unseen data is critical. Overfitting to specific datasets can limit the applicability of models to broader contexts.
- **Integration:** Combining data from multiple sources (multi-omics) and integrating diverse data types (e.g., genomic, proteomic, clinical) require sophisticated algorithms and computational infrastructure.

Machine learning and artificial intelligence are transforming biology by providing powerful tools to analyze complex data, uncover biological insights, and drive innovation in research and healthcare. As these technologies continue to advance, they will play an increasingly vital role in understanding biological systems, developing new therapies, and personalizing medicine, ultimately improving health outcomes and advancing our knowledge of life.

Computational Modeling and Simulations

Computational modeling and simulations are essential for understanding biological systems' dynamics and interactions. Models range from simple ordinary differential equations (ODEs) to complex partial differential equations (PDEs) that describe the spatial and temporal changes in biological processes. Simulations using these models allow researchers to explore hypotheses, predict outcomes, and design experiments.Computational modeling involves constructing mathematical or computational representations of biological systems. Simulations use these models to study the system's behavior under various conditions, providing insights that are often difficult or impossible to obtain experimentally.

Types of Models

- **Deterministic Models:** These models use fixed parameters and initial conditions to produce a single, predictable outcome. Ordinary differential equations (ODEs) and partial differential equations (PDEs) are common deterministic models used to describe continuous biological processes, such as enzyme kinetics or population dynamics.
- **Stochastic Models:** Stochastic models incorporate random variability, making them suitable for systems where intrinsic noise or randomness plays a significant role, such as gene expression and molecular interactions. Examples include stochastic differential equations and agent-based models.
- **Hybrid Models:** Hybrid models combine deterministic and stochastic approaches to capture both continuous processes and discrete events within a biological system. These models are useful for systems that exhibit both types of behaviors.

Techniques and Tools

Ordinary Differential Equations (ODEs)

ODEs are widely used to model dynamic biological systems, where the change in a system's state variables is described by differential equations. Applications include modeling the kinetics of biochemical reactions and population growth.

- **Michaelis-Menten Kinetics:** A classic example of ODEs in biology, this model describes the rate of enzymatic reactions based on substrate concentration and enzyme affinity.
- **Lotka-Volterra Equations:** These equations model predator-prey interactions and other ecological dynamics, illustrating how populations of different species interact over time.

Partial Differential Equations (PDEs)

PDEs extend ODEs to model spatial and temporal variations in biological systems, such as diffusion processes and pattern formation.

- **Reaction-Diffusion Models:** These models describe the diffusion of chemical substances in space and their reactions. They are used to study processes like morphogenesis and pattern formation in developmental biology.

Agent-Based Models

Agent-based models (ABMs) simulate the actions and interactions of individual agents (e.g., cells, organisms) to study emergent behaviors in a system. Each agent follows a set of rules, and the collective behavior of the system emerges from these local interactions.

- **Cellular Automata:** A type of ABM where the grid-based model represents individual cells, and each cell updates its state based on local rules. Applications include tissue growth and cancer modeling.

Network Models

Biological systems are often represented as networks, where nodes represent entities (genes, proteins) and edges represent interactions (regulatory, physical).

- **Gene Regulatory Networks (GRNs):** GRNs model the regulatory interactions between genes and transcription factors, providing insights into gene expression patterns and regulatory mechanisms.
- **Protein-Protein Interaction Networks (PPINs):** PPINs represent the physical interactions between proteins, helping to elucidate cellular processes and identify potential drug targets.

Applications

- **Cancer Modeling:** Computational models simulate tumor growth, progression, and response to treatment. For example, ODE and PDE models are used to study the dynamics of cancer cell populations and the effects of therapies.
- **Infectious Disease Spread:** Epidemiological models, such as the SIR (Susceptible-Infected-Recovered) model, use differential equations to predict the spread of infectious diseases and evaluate intervention strategies.

- **Pharmacokinetics and Pharmacodynamics (PK/PD):** Models describe how drugs are absorbed, distributed, metabolized, and excreted (pharmacokinetics) and their effects on the body (pharmacodynamics). These models inform dosing regimens and optimize drug efficacy.
- **Virtual Screening and Docking:** Computational simulations predict how small molecules interact with target proteins, aiding in the identification of potential drug candidates. Molecular dynamics simulations further refine these predictions by modeling the dynamic behavior of biomolecular complexes.
- **Metabolic Pathway Analysis:** Computational models of metabolic networks, using techniques like flux balance analysis (FBA), help understand metabolic fluxes and identify potential metabolic engineering targets.
- **Synthetic Biology:** Models guide the design of synthetic biological circuits and systems, predicting their behavior before experimental implementation.

Computational Tools and Software

- **MATLAB and Simulink:** Widely used for mathematical modeling and simulations, offering a range of toolboxes for biological applications.
- **COPASI (COmplexPAthwaySImulator):** A software application for simulation and analysis of biochemical networks and their dynamics.
- **CellDesigner:** A modeling tool for constructing and simulating biochemical networks, supporting both deterministic and stochastic simulations.
- **Virtual Cell (VCell):** A platform for modeling and simulation of cell biology processes, integrating multiple spatial and temporal scales.

Computational modeling and simulations are indispensable tools in computational biology, offering deep insights into the dynamics of biological systems. By combining mathematical rigor with computational power, these models enable researchers to predict complex behaviors, optimize therapeutic interventions, and advance our understanding of life. As computational methods and technologies continue to evolve, their applications in biology will become even more profound, driving innovation and improving health outcomes.

Data Retrieval and Management

Efficient data retrieval and management are critical for handling the vast amounts of biological data available. Tools and techniques such as SQL, data warehousing, and cloud computing are used to store, query, and manage biological databases. Proper data management ensures that data is accessible, reliable, and can be integrated across different studies and platforms.

Data Retrieval

- **Query Interfaces:** Most biological databases provide web-based query interfaces that allow users to search for data using keywords, accession numbers, or sequence similarity searches (e.g., BLAST).
- **APIs (Application Programming Interfaces):** APIs enable programmatic access to databases, allowing researchers to automate data retrieval and integrate database queries into their computational pipelines. For example, NCBI's E-utilities provide API access to various NCBI databases.

- **FTP (File Transfer Protocol):** Many databases offer bulk data downloads via FTP, facilitating large-scale data acquisition for local analysis.

Data Integration

- **Data Warehousing:** Integrating data from multiple databases into a single, centralized repository, or data warehouse, enables comprehensive analysis and comparison of diverse datasets.
- **Semantic Integration:** Using ontologies and standardized vocabularies to link data from different sources, ensuring consistency and facilitating cross-database queries. The Gene Ontology (GO) is a widely used resource for annotating genes and gene products with standardized terms.

Data Management

- **Database Management Systems (DBMS):** Systems like MySQL, PostgreSQL, and MongoDB are used to store, manage, and retrieve large volumes of biological data efficiently.
- **Data Curation:** The process of organizing, annotating, and validating data to ensure its accuracy, consistency, and usability. Many databases employ expert curators to maintain high-quality data.
- **Version Control:** Tracking changes in databases over time, allowing researchers to reference specific versions of data for reproducibility and comparison.

Biological databases are indispensable tools in modern biological research, providing vast amounts of data that drive discovery and innovation. By organizing and providing access to diverse types of biological information, these databases support a wide range of applications, from fundamental research to clinical practice. Efficient data retrieval, integration, and management practices are essential for maximizing the utility of these resources, enabling researchers to uncover new insights and advance our understanding of complex biological systems. As biological data continues to grow in volume and complexity, the development and maintenance of robust databases will remain a cornerstone of computational biology.

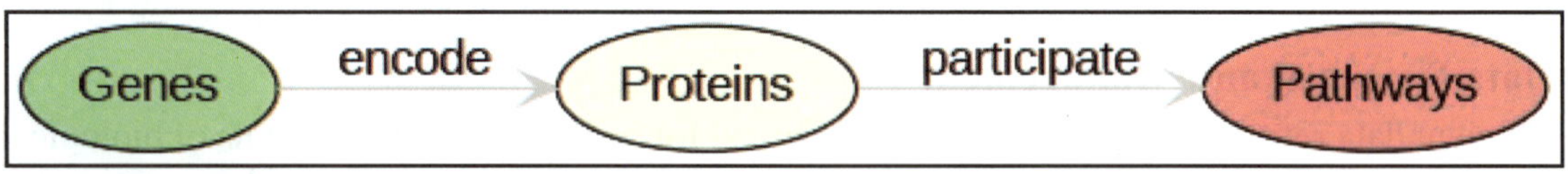

Figure 2: Illustrating the relationships between different entities in a biological database

Genomics and Proteomics

Genome Sequencing Technologies

Genome sequencing technologies have advanced rapidly, from the early Sanger sequencing to high-throughput next-generation sequencing (NGS). These technologies allow for the rapid and accurate sequencing of entire genomes, providing comprehensive insights into genetic variation and function.

- **Sanger Sequencing:** The first widely used method for DNA sequencing, known for its accuracy but limited throughput.

- **Next-Generation Sequencing (NGS):** High-throughput technologies, such as Illumina sequencing, that allow rapid sequencing of large genomes at a lower cost compared to Sanger sequencing.
- **Third-Generation Sequencing:** Technologies like PacBio and Oxford Nanopore offer long-read sequencing, which provides more comprehensive insights into genomic regions that are difficult to sequence with short-read technologies, such as structural variants and repetitive regions.

Sequence Alignment and Assembly

Sequence alignment algorithms, such as BLAST and ClustalW, are used to compare DNA or protein sequences to identify regions of similarity. Sequence assembly techniques, including de novo assembly and reference-guided assembly, reconstruct complete genomes from sequencing reads.

- **Sequence Alignment:** The process of arranging sequences of DNA, RNA, or protein to identify regions of similarity. Tools like BLAST (Basic Local Alignment Search Tool) and Clustal Omega are widely used for this purpose.
- **Genome Assembly:** Combining short DNA sequences into longer, continuous sequences to reconstruct the original genome. There are two main types of assembly:
- **De Novo Assembly:** Constructing genomes without a reference sequence, useful for studying novel organisms.
- **Reference-Based Assembly:** Aligning sequences to a known reference genome, facilitating the identification of variants.

Protein Structure Prediction and Analysis

Predicting and analyzing protein structures are critical for understanding their functions. Techniques such as homology modeling, molecular dynamics simulations, and AI-based methods like AlphaFold are employed to predict protein structures from amino acid sequences and study their interactions and stability.Proteomics is the large-scale study of proteins, including their structures, functions, and interactions. It aims to understand the proteome, the entire set of proteins produced by an organism, tissue, or cell.

Protein Sequencing and Identification

Mass Spectrometry (MS): The principal technology used in proteomics for identifying and quantifying proteins. MS measures the mass-to-charge ratio of ionized peptides and deduces protein identities based on peptide mass fingerprints and fragmentation patterns.

- **Edman Degradation:** A traditional method for sequencing amino acids in a protein, used less frequently today due to the efficiency of MS-based methods.

Protein Structure Prediction and Analysis

- **X-ray Crystallography:** A technique to determine the 3D structure of proteins by analyzing the diffraction pattern of X-rays passing through a crystallized protein.
- **Nuclear Magnetic Resonance (NMR) Spectroscopy:** Provides information about the physical and chemical properties of atoms in proteins, useful for studying protein dynamics in solution.

- **Cryo-Electron Microscopy (Cryo-EM):** An emerging method that allows the visualization of proteins and complexes at near-atomic resolution without the need for crystallization.
- **Computational Prediction:** Methods such as homology modeling, ab initio modeling, and AlphaFold, a revolutionary AI-based tool that predicts protein structures with high accuracy.

Applications in Proteomics

- Protein Expression Profiling: Quantifying protein expression levels across different conditions, tissues, or developmental stages to understand cellular functions and responses.
- **Post-Translational Modifications (PTMs):** Identifying chemical modifications of proteins, such as phosphorylation, glycosylation, and ubiquitination, which regulate protein function and activity.
- **Protein-Protein Interactions (PPIs):** Mapping the interaction networks between proteins to understand cellular processes and identify potential drug targets. Techniques include yeast two-hybrid screening, co-immunoprecipitation, and affinity purification-MS.
- **Biomarker Discovery:** Identifying proteins that serve as indicators of disease states, which can be used for diagnostics, prognostics, and therapeutic monitoring.

Applications in Genomics

- **Genomic Variation Analysis:** Identifying and interpreting genetic variants, such as single nucleotide polymorphisms (SNPs), insertions, deletions, and copy number variations, to understand their impact on phenotype and disease.
- **Comparative Genomics:** Comparing genomes from different species to study evolutionary relationships and identify conserved elements.
- **Functional Genomics:** Investigating the functions and interactions of genes and non-coding regions through techniques like RNA sequencing (RNA-seq), chromatin immunoprecipitation sequencing (ChIP-seq), and CRISPR-Cas9 screens.
- Personalized Medicine: Using individual genomic information to tailor medical treatments, predict disease risk, and develop personalized therapeutic strategies.

System Biology

Network Modeling and Analysis

Network modeling and analysis involve representing biological systems as networks of interacting components, such as genes, proteins, and metabolites. Tools like Cytoscape are used to visualize and analyze these networks, helping to identify key regulatory elements and interactions.

Types of Biological Networks

- **Gene Regulatory Networks (GRNs):** Represent the regulatory relationships between genes, where nodes represent genes and edges represent regulatory interactions (e.g., activation or repression).

- **Protein-Protein Interaction (PPI) Networks:** Depict physical interactions between proteins, highlighting how proteins collaborate to perform cellular functions.
- **Metabolic Networks:** Illustrate biochemical reactions within a cell, with nodes representing metabolites and edges representing enzymatic reactions.
- **Signaling Networks:** Show how signals are transmitted through a series of molecular interactions to regulate cellular processes.

Network Analysis Techniques

- **Topological Analysis:** Examines the structure of networks, including properties such as node degree, clustering coefficient, and network motifs, to identify key components and modules.
- **Pathway Analysis:** Identifies biological pathways within networks to understand the flow of information or metabolites and their impact on cellular functions.
- **Dynamic Modeling:** Uses mathematical models, such as ordinary differential equations (ODEs) and partial differential equations (PDEs), to simulate the temporal behavior of networks and predict their response to perturbations.

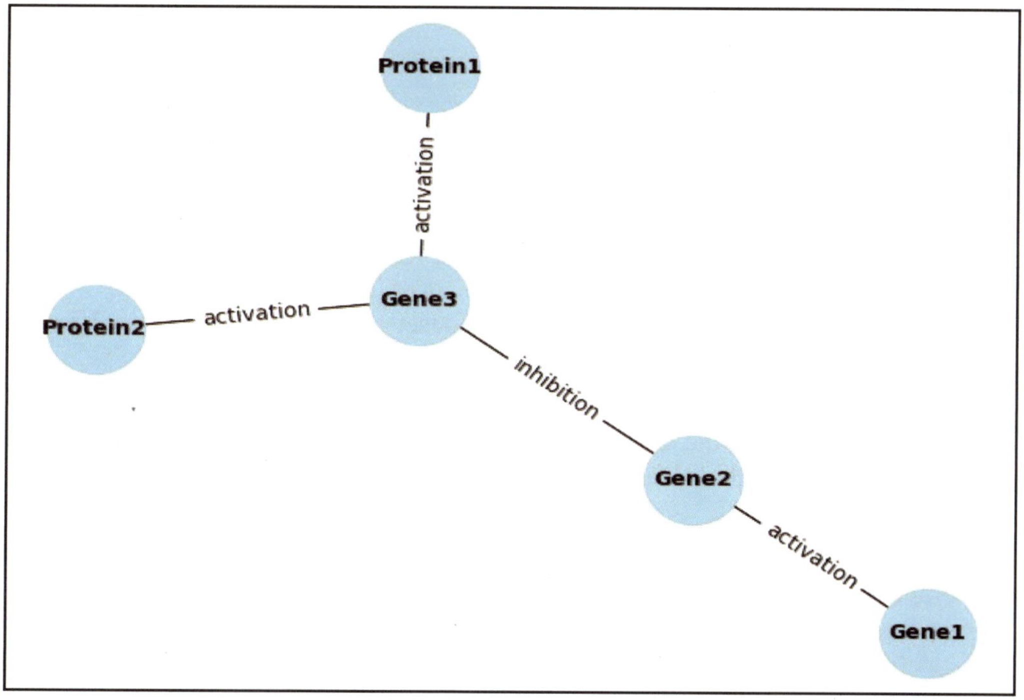

Figure 3:A network diagram can represent gene regulatory networks or protein-protein interaction networks, showing nodes as genes or proteins and edges as interactions

Pathway Reconstruction and Simulation

Pathway reconstruction involves identifying the biochemical pathways within a cell, while simulation models these pathways to understand their behavior under different conditions. Techniques such as flux balance analysis (FBA) are used to study metabolic pathways and predict cellular responses to environmental changes.

Methods for Pathway Reconstruction

- **Data-Driven Approaches:** Utilize high-throughput omics data to infer pathways. Techniques include correlation analysis, Bayesian networks, and machine learning algorithms.
- **Literature-Based Approaches:** Integrate information from scientific literature and curated databases (e.g., KEGG, Reactome) to construct detailed pathway maps.
- **Hybrid Approaches:** Combine data-driven and literature-based methods to improve the accuracy and completeness of pathway reconstructions.

Simulation Techniques

- **Deterministic Models:** Use differential equations to model the average behavior of pathway components over time, suitable for large populations of molecules.
- **Stochastic Models:** Incorporate random fluctuations to capture the inherent noise in biological systems, particularly important for systems with low molecule numbers.
- **Agent-Based Models:** Simulate the behavior of individual entities (e.g., cells) and their interactions within a system, useful for modeling heterogeneous populations and spatial dynamics.

Integration of Multiomics Data

Integrating multi-omics data (genomics, transcriptomics, proteomics, etc.) provides a holistic view of biological systems. Computational tools and pipelines are developed to combine these diverse data types, enabling comprehensive analyses that can reveal complex interactions and regulatory mechanisms.

Strategies for Multi-Omics Integration

- **Data Fusion:** Combining multiple omics datasets into a unified framework to identify correlations and causal relationships across different biological levels.
- **Hierarchical Modeling:** Building models that incorporate data from multiple omics layers, such as genomic data informing transcriptomic models, which in turn inform proteomic and metabolomic models.
- **Network Integration:** Constructing multi-layered networks that represent different types of interactions (e.g., gene-gene, protein-protein, metabolite-protein), facilitating comprehensive network analysis.

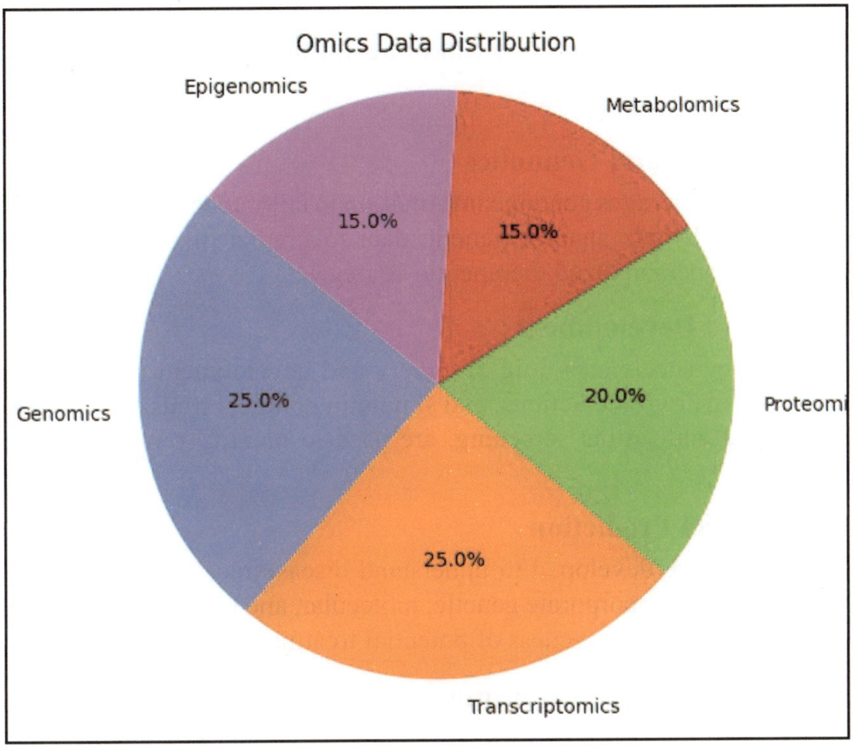

Figure 4: Represent the different omics data sets

Case Studies

Multiple Myeloma Systems Biology

In "ODE/PDE Analysis of Multiple Myeloma: Programming in R" by William E. Schiesser, systems biology approaches are used to model the progression of multiple myeloma, a type of bone cancer. Key points include:

- **Network Modeling:** Constructing gene regulatory and protein interaction networks to understand the molecular interactions driving the disease.
- **Pathway Simulation:** Using ODEs and PDEs to simulate the dynamic behavior of cancer pathways, predicting disease progression and response to treatments.
- **Multi-Omics Integration:** Combining genomic and proteomic data to identify key regulatory nodes and potential therapeutic targets.

Physical Oncology

"An Introduction to Physical Oncology" by Vittorio Cristini, Eugene J. Koay, and Zhihui Wang emphasizes the integration of physical and biological data to model cancer behavior. The book highlights:

- **Network Analysis:** Examining signaling and metabolic networks to understand tumor growth and metastasis.
- **Dynamic Modeling:** Using mathematical models to simulate tumor dynamics and predict responses to therapies.

- **Multi-Omics Integration:** Integrating genomic, transcriptomic, and proteomic data with physical models to optimize cancer treatment strategies.

Application in Medicine

Personalized Medicine and Genomics

Personalized medicine leverages genomic information to tailor medical treatments to individual patients. Computational tools analyze genetic data to predict disease risk, guide treatment decisions, and develop personalized therapeutic strategies.

Drug Discovery and Development

Computational biology accelerates drug discovery and development by identifying potential drug targets, predicting drug interactions, and simulating clinical trials. In silico methods, such as molecular docking and virtual screening, are used to identify promising compounds and optimize drug design.

Disease Modeling and Prediction

Computational models are developed to understand disease mechanisms and predict disease progression. These models incorporate genetic, molecular, and clinical data to simulate disease dynamics and evaluate the effectiveness of potential treatments.

Future Directions and Challenges

Emerging Technologies and Methodologies

Emerging technologies such as single-cell sequencing, CRISPR, and advanced imaging techniques are expanding the capabilities of computational biology. Methodologies like machine learning, artificial intelligence, and quantum computing promise to further enhance data analysis and model predictions.

Ethical Considerations in Computational Biology

The use of computational biology raises ethical considerations, particularly regarding data privacy, consent, and the potential for genetic discrimination. Ensuring ethical practices in data handling, analysis, and application is crucial to protect individuals' rights and foster public trust.

Future Trends and Research Directions

Future research in computational biology will likely focus on integrating multi-scale models, enhancing the precision of personalized medicine, and developing more robust and interpretable AI models. Collaborative efforts across disciplines will be essential to address complex biological questions and improve health outcomes.

Conclusion

Computational biology is a rapidly evolving field that intersects biology, computer science, mathematics, and statistics, providing essential tools and methodologies for understanding complex biological systems from the molecular level to entire ecosystems. This book chapter has explored foundational concepts, techniques, and applications of computational biology, emphasizing its interdisciplinary importance. Key points include the historical roots in early bioinformatics, significant advances through computer technology and data science, and core concepts like algorithms, data structures, machine learning, AI, computational modeling, and simulations. Biological databases like GenBank, PDB, and UniProt are crucial for research, while advanced sequencing technologies and bioinformatics tools have revolutionized genomics and proteomics. Systems biology integrates network modeling, pathway reconstruction, and multi-omics data to study complex interactions and disease mechanisms. Applications in medicine highlight the role of computational biology in personalized medicine, drug discovery, and disease modeling, driven by emerging technologies like single-cell sequencing and CRISPR. The future of computational biology is promising, with advancements in high-throughput sequencing, imaging technologies, and AI, emphasizing integration and interdisciplinarity. Ethical considerations, data privacy, and equitable access are critical challenges that researchers and policymakers must address. As a cornerstone of modern biological research, computational biology continues to drive innovations that enhance human health and our understanding of the natural world.

References

1. Tabula Muris Consortium; Overall coordination; Logistical coordination; Organ collection and processing; Library preparation and sequencing; Computational data analysis; Cell type annotation; Writing group; Supplemental text writing group; Principal investigators. Single-cell transcriptomics of 20 mouse organs creates a Tabula Muris. Nature. 2018 Oct;562(7727):367-372. doi: 10.1038/s41586-018-0590-4. Epub 2018 Oct 3. PMID: 30283141; PMCID: PMC6642641.

2. Tabula Muris Consortium; Overall coordination; Logistical coordination; Organ collection and processing; Library preparation and sequencing; Computational data analysis; Cell type annotation; Writing group; Supplemental text writing group; Principal investigators. Single-cell transcriptomics of 20 mouse organs creates a Tabula Muris. Nature. 2018 Oct;562(7727):367-372. doi: 10.1038/s41586-018-0590-4. Epub 2018 Oct 3. PMID: 30283141; PMCID: PMC6642641.

3. Pankow S, Martínez-Bartolomé S, Bamberger C, Yates JR. Understanding molecular mechanisms of disease through spatial proteomics. CurrOpinChem Biol. 2019 Feb;48:19-25. doi: 10.1016/j.cbpa.2018.09.016. Epub 2018 Oct 9. PMID: 30308467; PMCID: PMC6435883.

4. Fischer SC. An Introduction to Image-Based Systems Biology of Multicellular Spheroids for Experimentalists and Theoreticians. In: Husi H, editor. Computational Biology [Internet]. Brisbane (AU): Codon Publications; 2019 Nov 21. Chapter 1. Available from: https://www.ncbi.nlm.nih.gov/books/NBK550340/ doi: 10.15586/computationalbiology.2019.ch1

5. Urbanski AH, Araujo JD, Creighton R, et al. Integrative Biology Approaches Applied to Human Diseases. In: Husi H, editor. Computational Biology [Internet]. Brisbane (AU): Codon Publications; 2019 Nov 21. Chapter 2. Available from: https://www.ncbi.nlm.nih.gov/books/NBK550336/ doi: 10.15586/computationalbiology.2019.ch2

6. Sophie Bekisz, Liesbet Geris,Cancer modeling: From mechanistic to data-driven approaches, and from fundamental insights to clinical applications,Journal of Computational Science,Volume 46,2020,101198,ISSN 1877-7503,https://doi.org/10.1016/j.jocs.2020.101198.

7. Harkos C, Stylianopoulos T, Jain RK. Mathematical modeling of intratumoral immunotherapy yields strategies to improve the treatment outcomes. PLoSComput Biol. 2023 Dec 19;19(12):e1011740. doi: 10.1371/journal.pcbi.1011740. PMID: 38113269; PMCID: PMC10763956.

6 Biological Database:Unveiling The Digital Atlas of Life

Biological databases represent the cornerstone of modern biological research, serving as reservoirs of vast and intricate biological data critical for deciphering the complexities of life. This chapter embarks on a comprehensive exploration of these databases, elucidating their diverse types, pivotal functionalities, inherent challenges, and promising future trajectories within the realm of biological sciences. Biological databases serve as repositories for diverse biological data, from DNA sequences to protein structures, enabling researchers to access, analyze, and interpret information crucial for understanding life's complexities. Their evolution from early repositories to sophisticated integrated systems has revolutionized biological research.

Introduction

Biological databases represent digital archives meticulously curated to house diverse biological data. They encompass an array of information categories, including genomic sequences, protein structures and functions, metabolic pathways, scientific literature, and ontologies defining biological concepts. These repositories serve as indispensable tools, facilitating comprehensive access to biological data that are critical for scientific inquiry, innovation, and practical applications across various domains within the life sciences. Whether unraveling the genomic sequences of organisms, deciphering protein-protein interactions, or exploring the intricacies of cellular metabolism, biological databases provide researchers with an unprecedented resource to delve deeper into the complexities of life. These databases not only store information but also serve as catalysts for scientific discoveries, enabling researchers to navigate the intricate landscape of biological data, unveil novel insights, and fuel transformative innovations in medicine, agriculture, ecology, and beyond.

Need of Biological Data

The need for biological data is fundamental to advancing our understanding of life at its most intricate levels. Biological data forms the bedrock of scientific exploration and comprehension in life sciences, serving as the fundamental currency that drives research, innovation, and transformative discoveries. At its essence, the need for biological data arises from the intricate and multifaceted nature of living organisms, encompassing their genetic makeup, molecular interactions, physiological responses, and ecological dynamics. Several key factors underscore the critical need for biological data:

Unraveling Life's Complexity

Life, in all its forms, exhibits remarkable complexity. Biological data—ranging from genomic sequences to protein structures, metabolic pathways, and ecological interactions—serves as a window into this complexity. It allows scientists to unravel the intricacies of genetic codes, cellular mechanisms, and the interplay of biological systems, providing invaluable insights into the fundamental principles governing life.

Driving Scientific Inquiry and Innovation

Biological data fuels scientific inquiry across diverse domains. It acts as the fuel for hypothesis generation, experimental design, and the validation of scientific theories. These data empower researchers to explore new frontiers, develop novel treatments, devise innovative technologies, and address pressing challenges such as disease, environmental degradation, and food security.

Facilitating Precision and Personalized Medicine

In the realm of medicine, biological data plays a pivotal role in advancing precision and personalized healthcare. Genomic data, for instance, enables the identification of disease-associated genetic variants, facilitating tailored therapies and personalized treatment strategies. The integration of diverse biological datasets enhances diagnostic accuracy and treatment efficacy, ushering in a new era of precision medicine.

Informing Conservation and Environmental Management

Biological data aids in understanding ecosystems, species diversity, and environmental dynamics. It informs conservation efforts by providing insights into endangered species, ecological networks, and the impact of human activities on biodiversity. Such data-driven insights are crucial for formulating effective conservation strategies and sustainable environmental management practices.

Accelerating Agricultural and Food Sciences

In agriculture, biological data contributes significantly to crop improvement, pest resistance, and sustainable agricultural practices. Understanding plant genetics, soil microbiomes, and environmental factors through biological data aids in developing resilient crop varieties, ensuring food security, and mitigating agricultural challenges.

Fueling Technological Advancements

Biological data serves as the foundation for technological advancements. From bioinformatics tools facilitating data analysis to cutting-edge technologies like CRISPR-Cas9 derived from biological insights, these innovations are transforming industries, enabling breakthroughs in biotechnology, pharmaceuticals, and beyond.

In essence, the need for biological data transcends scientific curiosity; it is an indispensable catalyst propelling humanity towards a deeper understanding of life's intricacies. Its application spans a spectrum of disciplines, paving the way for innovative solutions to global challenges and fostering a more profound appreciation of the natural world.

Unraveling the Evolution of Biological Databases: A Journey Through Time

Genesis: The Dawn of Data Repositories

The roots of biological databases trace back to the mid-20th century when the scientific community recognized the need to systematically organize biological information. One of the pioneering repositories emerged with the establishment of GenBank in the early 1970s, heralding the era of sequence databases. GenBank, initially a modest collection of genetic sequences, rapidly expanded as DNA sequencing technologies advanced, becoming a cornerstone resource for genetic information worldwide.

The Genomic Era: Unveiling the Code of Life

The unveiling of the human genome in the early 2000s marked a pivotal juncture in biological database history. Projects like the Human Genome Project catalyzed the exponential growth of genomic databases. Institutions worldwide collaborated to construct comprehensive genomic repositories like Ensembl, UCSC Genome Browser, and the National Center for Biotechnology Information (NCBI) databases, laying the foundation for understanding genetic diversity across species.

Proteomics and Beyond: Expanding Horizons

As scientific exploration extended beyond genes, the focus shifted to proteins and their functions. Protein Data Bank (PDB) emerged as a pioneering repository housing three-dimensional structures of proteins, essential for drug design, molecular modeling, and understanding protein interactions. Simultaneously, other repositories like UniProt and InterPro flourished, cataloging protein sequences, functions, and domains, enriching the understanding of molecular biology.

Metabolic Pathways and Integrative Databases: Holistic Insights

The integration of diverse data types became imperative to unravel the complexities of biological systems. Metabolic databases such as KEGG and Reactome offered comprehensive insights into biochemical pathways, metabolic networks, and their associations with diseases. This era also witnessed the emergence of integrative databases like the Gene Expression Omnibus (GEO), facilitating the aggregation of diverse biological data—gene expression profiles, genomic variations, and clinical information—for comprehensive analysis.

Ontologies and Standardization: Building Semantic Bridges

A critical milestone emerged with the advent of ontologies and standardized vocabularies. Efforts like the Gene Ontology (GO) provided a structured framework for annotating gene functions, cellular components, and biological processes. These ontologies not only facilitated data standardization but also enabled interoperability and semantic integration across disparate databases.

Challenges and Innovations: Navigating the Data Deluge

The evolution of biological databases was not devoid of challenges. With the exponential growth of data, challenges encompassing data integration, interoperability, data quality, and ethical considerations emerged. To address these, innovations incorporating artificial intelligence, machine learning, and blockchain technology are revolutionizing data curation, analysis, and security, propelling databases toward enhanced efficiency and reliability.

The Future Horizon: Towards Comprehensive Insights

As we stride into the future, biological databases continue to evolve, promising deeper insights into the intricacies of life. Efforts are underway to forge predictive databases, personalized medicine platforms, and community-driven curation models. Collaborative endeavors and interdisciplinary approaches will shape the next chapter, unraveling the enigmas of biology and propelling humanity toward transformative discoveries.

Exploring the Hierarchy: Primary, Secondary, and Derived Databases in Biological Sciences

Primary Databases: Foundations of Raw Data

Primary databases serve as repositories of raw, unprocessed data obtained directly from experimental assays and observations. These databases capture fundamental biological information at its source, offering the initial groundwork for subsequent analyses and interpretations. Primary databases lay the foundation for scientific discoveries and form the backbone of biological research.

Key Features of Primary Databases

Raw Data Repository: Primary databases house original data obtained from experiments, sequencing, imaging, and other raw sources.

Examples: GenBank for DNA sequences, Protein Data Bank (PDB) for protein structures, and Sequence Read Archive (SRA) for raw sequencing data.

Secondary Databases: Curated and Annotated Information

Secondary databases process, curate, and annotate raw data obtained from primary sources, offering structured and annotated information. These databases add value to raw data by organizing and standardizing it, making it more accessible and interpretable for researchers across various disciplines within biological sciences.

Key Features of Secondary Databases

Curated Information: Secondary databases compile and curate data obtained from primary sources, providing standardized annotations and classifications.

Examples: UniProt for protein sequences, RefSeq for curated genomic sequences, and Model Organism Databases (MODs) for model organisms.

Derived Databases: Synthesized Insights and Analytical Outputs

Derived databases represent higher-level repositories that integrate, analyze, and synthesize information obtained from primary and secondary sources. These databases offer processed data, analytical outputs, and derived knowledge, providing a holistic view and interpretations essential for complex biological inquiries and systems-level understanding.

Key Features of Derived Databases

Analytical Outputs: Derived databases offer processed data, computational models, and analytical outputs from integrated datasets.

Examples: KEGG for metabolic pathways, STRING for protein-protein interactions, and The Cancer Genome Atlas (TCGA) for cancer genomics.

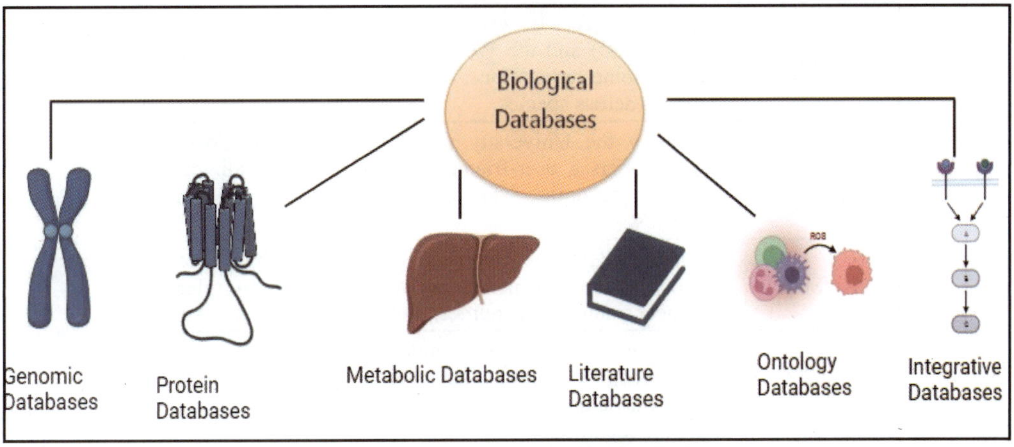

Figure 1: Types of Biological Databases

BIOLOGICAL DATABASE

Genomic Databases: Deciphering Genetic Blueprints

At the heart of biological exploration lie genomic databases, repositories that store genetic sequences, genomes, and variations across species. GenBank, a pioneering database, paved the way for housing nucleotide sequences, followed by databases like Ensembl and UCSC Genome Browser, offering comprehensive genome annotations and visualization tools. These repositories provide researchers with a digital atlas to explore the genetic codes that underpin life's diversity.

Key Features of Genomic Databases

a. **Sequence Repositories:** Genomic databases initially emerged as repositories for DNA and RNA sequences. They encompass a vast array of sequences, including whole genomes, gene-specific sequences, and transcriptomes.

b. **Annotations and Annotations:** These databases offer annotations, providing valuable information on gene locations, structures, regulatory elements, and variations, enabling researchers to decipher functional elements within genomes.

c. **Comparative Genomics:** Comparative genomics sections within these databases allow the comparison of genetic sequences across different species, shedding light on evolutionary relationships, genetic conservation, and divergence.

d. **Visualization Tools:** Many genomic databases provide visualization tools, allowing researchers to explore and analyze genomic data through interactive interfaces, genome browsers, and alignment viewers.

Table 1: Major Genomic Biological Databases

Genomic Databases	Description
GenBank	Established by the National Center for Biotechnology Information (NCBI), GenBank stands as one of the oldest and most comprehensive repositories of genetic sequences from various organisms

Ensembl	A joint project between the European Bioinformatics Institute (EBI) and the Wellcome Sanger Institute, Ensembl offers genome annotations, variation data, and comparative genomics across species
UCSC Genome Browser	Hosted by the University of California, Santa Cruz, this browser offers a user-friendly interface for visualizing and exploring genomic data, including genomes of various organisms
RefSeq	Curated by NCBI, the Reference Sequence Database (RefSeq) provides a comprehensive, curated collection of reference sequences, including genomes, transcripts, and proteins
dbSNP	Focused on variations, the Single Nucleotide Polymorphism Database (dbSNP) provides information on variations in DNA sequences, aiding in understanding genetic diversity and disease associations
1000 Genomes Project	This project cataloged human genetic variations from diverse populations, offering insights into human genetic diversity and disease susceptibility
The Genome Aggregation Database (gnomAD)	Focused on aggregating and harmonizing exome and genome sequencing data from diverse populations, gnomAD aids in understanding rare genetic variations
FlyBase	Catering specifically to Drosophila researchers, FlyBase contains genomic and genetic information on fruit flies, facilitating studies on developmental biology and genetics
The Arabidopsis Information Resource (TAIR)	Focused on the model plant Arabidopsis thaliana, TAIR offers genomic information essential for plant biology and genetics

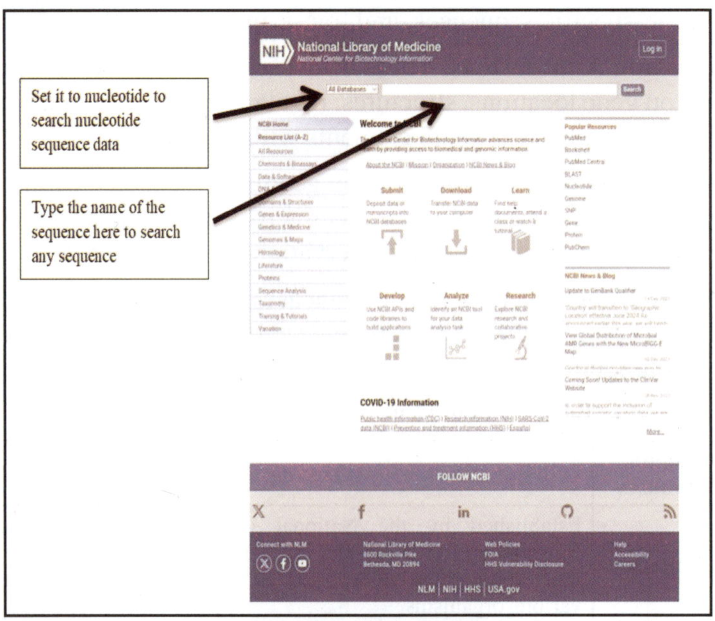

Figure 2: Homepage of NCBI

Figure 3: Searching Insulin Sequence in NCBI in GenBank Format

Protein Databases: Unveiling Molecular Machinery

Complementing genomic repositories, protein databases catalog sequences, structures, functions, and interactions of proteins. The Protein Data Bank (PDB) stands as a cornerstone, offering three-dimensional structures critical for understanding protein folding, function, and drug design. UniProt, another pivotal database, aggregates protein sequences and functional information, aiding in protein characterization and annotation.

Key Features of Protein Databases

a. **Protein Sequences:** These databases encompass vast collections of protein sequences derived from experimental and computational methods, providing a comprehensive catalogue of known proteins.

b. **Protein Structures:** Integral to protein databases are three-dimensional structures, elucidating the spatial arrangement of amino acids, facilitating insights into protein folding, active sites, and interactions with other molecules.

c. **Functional Annotations:** Annotations within protein databases offer information about protein function, domains, motifs, and post-translational modifications, aiding in understanding the biological roles of proteins.

d. **Protein-Protein Interactions:** Many databases compile data on protein-protein interactions, delineating networks of molecular interactions crucial for cellular processes.

Table 2: Major Protein Biological Databases

Protein Databases	Description
Protein Data Bank (PDB)	PDB stands as a seminal repository housing experimentally determined three-dimensional structures of proteins, nucleic acids, and complexes, providing a wealth of structural information crucial for drug design and structural biology
UniProt	UniProt is a comprehensive resource offering protein sequences and functional information, amalgamating data from various sources to provide an extensive and annotated protein database
InterPro	This database integrates diverse information about protein families, domains, and functional sites, using predictive models to categorize and classify protein sequences into families and domains
Pfam	Pfam focuses on protein families, offering curated collections of protein domains and families, aiding in functional annotations and evolutionary studies
Protein Information Resource (PIR)	PIR curates and annotates protein sequences, offering comprehensive protein information, including functional domains, classifications, and annotations
Structural Classification of Proteins (SCOP) and CATH	These databases classify protein structures into hierarchies, providing insights into evolutionary relationships and structural similarities among proteins
STRING	Focused on protein-protein interactions, STRING integrates experimental and predicted interaction data, facilitating the exploration of protein networks and functional associations
SWISS-MODEL	SWISS-MODEL offers tools and resources for protein structure modeling and prediction, aiding researchers in studying protein structures computationally
The Universal Protein Resource (UniProtKB):	UniProtKB offers a comprehensive collection of protein sequences, providing detailed information on protein functions, interactions, and annotations

Figure 4: Homepage of RCSB PDB Database

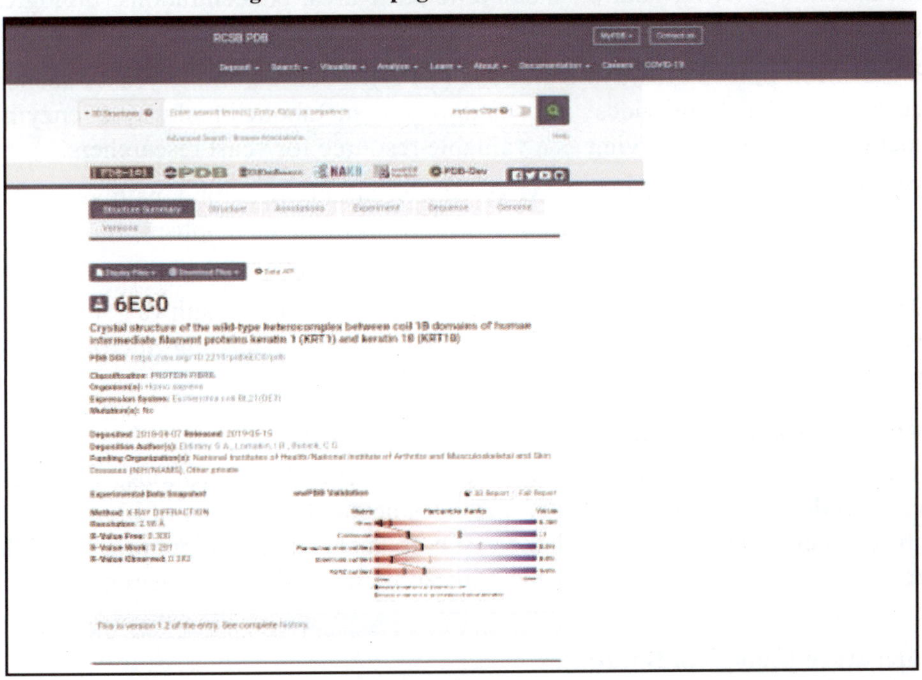

Figure 5: Searching Keratin Protein Structure in PDB Database

Delving into the intricate world of metabolism, these databases capture metabolic pathways, enzymes, and metabolites, offering insights into biochemical reactions within cells. KEGG

(Kyoto Encyclopedia of Genes and Genomes) and Reactome serve as encyclopedic resources, delineating metabolic networks, signaling pathways, and their associations with diseases, paving the way for understanding physiological processes and drug development.

Major Metabolic Biological Databases

a. **Kyoto Encyclopedia of Genes and Genomes (KEGG):** KEGG stands as a comprehensive resource offering information on metabolic pathways, signaling networks, and disease-related pathways across various organisms, providing valuable insights into biological systems.

b. **Reactome:** Reactome focuses on human biological pathways, offering curated information on reactions, pathways, and interactions, facilitating an understanding of human metabolism and its associations with diseases.

c. **BioCyc:** BioCyc houses a collection of pathway/genome databases (PGDBs) covering various organisms, providing curated information on metabolic pathways, regulatory networks, and genome-based predictions.

d. **MetaCyc:** MetaCyc serves as a curated database containing experimentally elucidated metabolic pathways and enzymes across diverse organisms, aiding in comparative and evolutionary studies.

e. **Human Metabolome Database (HMDB):** HMDB compiles information on human metabolites, offering data on metabolite structures, concentrations, biological roles, and associations with diseases.

f. **Yeast Metabolic Pathways Database (YPD):** Focused on the yeast Saccharomyces cerevisiae, YPD provides information on metabolic pathways, enzymes, and metabolites in yeast, serving as a valuable resource for yeast researchers.

g. **Plant Metabolic Network (PMN):** PMN offers metabolic pathway information specifically for plants, facilitating research on plant metabolism, biochemistry, and agricultural biotechnology.

h. **MicrobesOnline:** MicrobesOnline contains metabolic pathway information for various microbial species, aiding researchers in understanding microbial physiology and biochemistry

Literature Databases: Curating Scientific Knowledge

Beyond molecular data, literature databases curate scientific publications, annotations, and references. PubMed, a seminal repository, provides access to millions of biomedical articles, enabling researchers to navigate through the vast corpus of scientific literature. These databases act as treasure troves of knowledge, facilitating literature-based research and enhancing scientific discoveries.

Major Literature Biological Databases

a. **PubMed:** Operated by the National Center for Biotechnology Information (NCBI), PubMed stands as a vast repository of biomedical literature, offering access to millions of citations and abstracts from diverse scientific journals.

b. **Google Scholar:** Google Scholar provides a broad search platform spanning various disciplines, offering access to scholarly articles, theses, books, conference papers, and patents.

c. **Web of Science:** This database offers comprehensive coverage of scientific literature across disciplines, providing access to citation databases, enabling researchers to explore citation networks and impact factors.

d. **Scopus:** Scopus offers access to a wide range of scientific literature, providing citation analysis, abstracts, and references, aiding researchers in tracking scholarly publications and collaborations.

Ontologies and Knowledge Bases: Standardizing Biological Concepts

Ensuring coherence and semantic integration across diverse databases are ontologies and knowledge bases. The Gene Ontology (GO) and Human Phenotype Ontology (HPO) offer standardized vocabularies, categorizing gene functions, cellular components, and phenotypic traits. These structured frameworks foster uniformity, aiding in data interpretation and facilitating interoperability between databases.

Major Ontologies and Knowledge Bases

a. **Gene Ontology (GO):** GO stands as one of the most influential ontologies, categorizing genes and gene product attributes across species into biological processes, molecular functions, and cellular components.

b. **Human Phenotype Ontology (HPO):** Focused on human diseases and phenotypes, HPO provides a standardized vocabulary describing human phenotypic abnormalities linked to genetic conditions.

c. **Medical Subject Headings (MeSH):** Curated by the National Library of Medicine, MeSH categorizes biomedical literature and information into standardized terms, aiding in literature retrieval and indexing.

d. **The Plant Ontology (PO):** PO offers a standardized vocabulary for plant structures and growth stages, facilitating research in plant biology, genetics, and crop improvement.

e. **ChEBI (Chemical Entities of Biological Interest):** ChEBI categorizes and annotates chemical entities, providing structured information about molecular entities, helping in understanding chemical biology.

f. **The Cell Ontology (CL):** CL focuses on cellular components, providing a hierarchical representation of cell types and cellular structures, aiding in cell biology and pathology studies.

g. **The Environment Ontology (ENVO):** ENVO categorizes environmental features, habitats, and ecosystems, facilitating research in ecology, environmental sciences, and biodiversity studies.

Integrative Databases: Synthesizing Diverse Data Streams

In an era of data integration, these databases amalgamate diverse biological data types—genomic, proteomic, and clinical information—providing comprehensive insights. The Gene Expression Omnibus (GEO) integrates gene expression profiles, enabling researchers to explore gene regulation patterns across various conditions. Such integrative databases empower researchers with holistic perspectives and facilitate cross-disciplinary analyses.

Major Integrative Biological Databases

a. **The Gene Expression Omnibus (GEO):** GEO integrates gene expression profiles, offering a repository for high-throughput functional genomic data across diverse organisms and experimental conditions.

b. **The Cancer Genome Atlas (TCGA):** TCGA consolidates genomic, transcriptomic, and epigenomic data from various cancer types, enabling comprehensive analyses for understanding cancer biology.

c. **The ENCODE Project:** Focused on the functional annotation of the human genome, ENCODE integrates diverse genomic datasets, including DNA sequences, regulatory elements, and chromatin structure.

d. **The Human Protein Atlas (HPA):** HPA synthesizes protein expression data and spatial information across human tissues and organs, providing insights into human protein profiles and localization.

e. **STRING Database:** STRING consolidates protein-protein interaction data, combining experimental evidence and predicted interactions across species, facilitating the exploration of protein networks.

f. **The Cancer Cell Line Encyclopedia (CCLE):** CCLE integrates genomic and pharmacological data from a vast collection of cancer cell lines, aiding in understanding drug responses and cancer vulnerabilities.

g. **The European Nucleotide Archive (ENA):** ENA serves as a comprehensive repository for nucleotide sequence data, aggregating data from various sources and offering tools for data analysis and visualization.

h. **The Model Organism Databases (MODs):** MODs integrate genetic, genomic, and functional data for model organisms like Drosophila, Caenorhabditis elegans, and yeast, supporting research in genetics and developmental biology.

Navigating the Depths: Data Mining in Biological Databases

In the realm of biological sciences, the wealth of data available in biological databases has transformed research, enabling scientists to unravel the complexities of life's fundamental processes. Data mining, a powerful tool in this domain, involves extracting patterns, knowledge, and insights from vast biological datasets. This article explores the significance, techniques, and applications of data mining in biological databases, shaping the landscape of modern biology.

The Significance of Data Mining in Biological Databases

Biological databases store an extensive array of genetic, molecular, and clinical information, offering a treasure trove for researchers. However, the sheer volume and complexity of this data necessitate advanced computational techniques to unlock its potential. Data mining techniques act as a guiding light, helping researchers navigate this labyrinth of information, enabling:

a. **Pattern Discovery:** Data mining uncovers hidden patterns within biological datasets, revealing relationships between genes, proteins, pathways, and phenotypes. It assists in identifying biomarkers, regulatory motifs, and associations crucial for understanding biological systems.

b. **Predictive Modeling:** Machine learning algorithms in data mining predict biological outcomes, such as disease susceptibility, drug responses, or protein functions. These models aid in diagnostics, drug discovery, and personalized medicine.

c. **Integration of Multi-Omics Data:** With the advent of multi-omics datasets (genomics, transcriptomics, proteomics), data mining enables the integration of diverse data types, fostering a comprehensive understanding of biological interactions and networks.

d. **Knowledge Generation:** By mining biological databases, new biological hypotheses and insights emerge, leading to novel discoveries and scientific advancements in fields like genetics, molecular biology, and pharmacology.

Techniques Used in Data Mining for Biological Databases

a. **Clustering and Classification:** Clustering algorithms group similar biological entities, aiding in categorizing genes, proteins, or diseases based on common characteristics. Classification techniques predict biological outcomes, such as disease classifications or protein functions.

b. **Association Rule Mining:** This technique identifies relationships and associations between biological entities, revealing dependencies and co-occurrences, such as gene interactions or drug-target relationships.

c. **Dimensionality Reduction:** To handle high-dimensional biological data, techniques like principal component analysis (PCA) or t-distributed stochastic neighbor embedding (t-SNE) reduce data complexity while preserving essential information.

d. **Deep Learning:** Neural networks and deep learning architectures extract intricate patterns from biological data, enabling tasks like image analysis, sequence prediction, and drug discovery.

Applications of Data Mining in Biological Databases

a. **Drug Discovery and Development:** Data mining aids in virtual screening, predicting drug-target interactions, and identifying potential drug candidates from vast chemical libraries.

b. **Precision Medicine:** By analyzing genetic and clinical data, data mining enables personalized treatment strategies, tailoring therapies based on individual genetic profiles.

c. **Biomarker Discovery:** Mining biological databases helps identify biomarkers for diseases, aiding in early diagnosis and prognosis prediction.

d. **Systems Biology:** Integrating multi-omics data through data mining contributes to understanding complex biological systems and networks.

The Human Genome Project: Decoding the Blueprint of Humanity

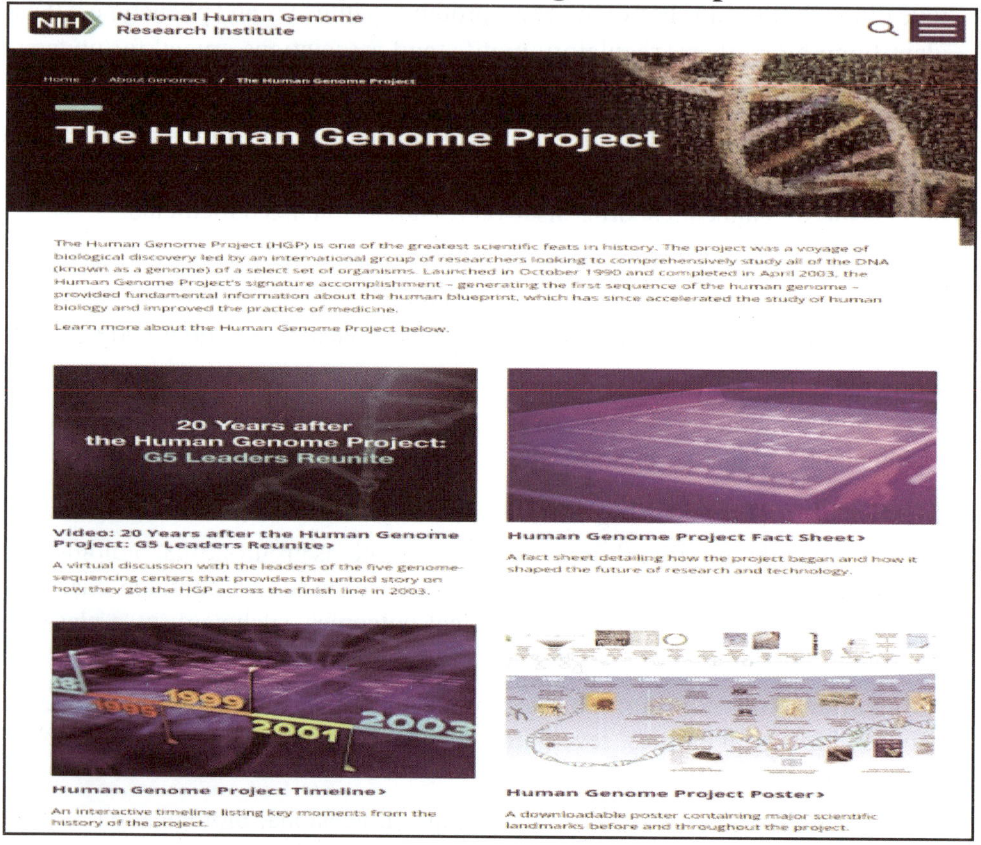

Figure 6: Human Genome Project

The Human Genome Project (HGP) stands as one of the most groundbreaking scientific endeavors in history, marking a pivotal milestone in our understanding of human biology. This ambitious international effort aimed to sequence and map the entire human genome, unraveling the intricate code that defines our genetic makeup. This article delves into the history, significance, achievements, and lasting impact of the Human Genome Project.

Origins and Goals of the Human Genome Project

In 1990, the HGP emerged as a collaborative effort involving researchers from across the globe, spurred by the vision to decode the human genome—a monumental task that was projected to revolutionize medicine, biology, and our comprehension of human life. The primary objectives included:

a. **Sequencing the Human Genome:** The central goal was to determine the complete sequence of DNA base pairs in the human genome, comprising over three billion nucleotides across 23 chromosome pairs.

b. **Annotating Genes and Identifying Variants:** Beyond sequencing, the project aimed to annotate genes, identify regulatory elements, and catalogue genetic variations, laying the foundation for understanding genetic diversity and disease susceptibility.

Scientific Breakthroughs and Achievements

The Human Genome Project progressed through several phases, harnessing technological advancements and collaborative efforts to achieve its monumental goals. Key milestones include:

a. **Draft Sequence Completion:** In 2000, the initial draft sequence of the human genome was published, marking a significant milestone in deciphering our genetic code. This draft provided a rough framework for further refinement and analysis.

b. **Completion of the Human Genome Sequence:** By 2003, the HGP reached its climax, culminating in the publication of the complete and highly accurate human genome sequence—a remarkable feat achieved ahead of schedule.

c. **Insights into Genetic Diseases:** The project led to the identification of genes associated with various diseases, offering insights into genetic predispositions and paving the way for targeted therapies and diagnostic tests.

d. **Technological Advancements and Data Sharing:** The HGP spurred technological innovations in DNA sequencing, leading to cost reductions and faster sequencing methods. Furthermore, it established principles of open data sharing, fostering collaborative research and accelerating scientific discoveries.

These repositories, teeming with genomic sequences, protein structures, metabolic pathways, and a myriad of biological information, stand as the bedrock of modern scientific inquiry. They are more than repositories; they are gateways to discovery, offering insights into the intricate mechanisms governing life itself. Navigating these databases reveals a world where raw data transforms into curated annotations and derived knowledge. They fuel research in genetics, molecular biology, and beyond, unlocking secrets of disease mechanisms, guiding drug development, and enabling personalized medicine. The creation of these databases is a testament to collaborative endeavors, uniting scientists, institutions, and nations in a shared pursuit of knowledge. Yet, this landscape is not static; it evolves, expands, and integrates with each technological leap and scientific advance.

However, their power also raises ethical considerations. The responsible handling of sensitive genetic and medical information is crucial, highlighting the importance of ethical, legal, and social considerations in their use and dissemination.

In essence, these databases encapsulate humanity's relentless quest to understand life's intricacies. They hold not just genetic blueprints but also the collective efforts, dreams, and aspirations of the scientific community. They promise a future where biological mysteries unravel, propelling us toward innovations that shape our understanding of life itself.

References

1. Zou D, Ma L, Yu J, Zhang Z. Biological databases for human research. Genomics Proteomics Bioinformatics. 2015 Feb;13(1):55-63. doi: 10.1016/j.gpb.2015.01.006. Epub 2015 Feb 21. PMID: 25712261; PMCID: PMC4411498.

2. Introduction to Biological Databases, Essential Bioinformatics , pp. 10 – 28, Publisher: Cambridge University Press, Print publication year: 2006, DOI: https://doi.org/10.1017/CBO9780511806087.003

3. Lina Ma, Dong Zou, Lin Liu, Huma Shireen, Amir A. Abbasi, Alex Bateman, Jingfa Xiao, Wenming Zhao, Yiming Bao, Zhang Zhang, Database Commons: A Catalog of Worldwide Biological Databases, Genomics, Proteomics & Bioinformatics,Volume 21, Issue 5, 2023,Pages 1054-1058,ISSN 1672-0229, https://doi.org/10.1016/j.gpb.2022.12.004

4. Robbins, R.J. Biological databases: A new scientific literature. Publishing Research Quarterly 10, 3–27 (1994). https://doi.org/10.1007/BF02680434

5. Singh, G.B. (2015). Biological Databases. In: Fundamentals of Bioinformatics and Computational Biology. Modeling and Optimization in Science and Technologies, vol 6. Springer, Cham. https://doi.org/10.1007/978-3-319-11403-3_3

6. Yang B, Mao J, Gao B, Lu X. Computer-Assisted Drug Virtual Screening Based on the Natural Product Databases. Curr Pharm Biotechnol. 2019;20(4):293-301. doi: 10.2174/1389201020666190328115411. PMID: 30919773.

7. Sunita, Sajid A, Singh Y, Shukla P. Computational tools for modern vaccine development. Hum Vaccin Immunother. 2020 Mar 3;16(3):723-735. doi: 10.1080/21645515.2019.1670035. Epub 2019 Dec 18. PMID: 31545127; PMCID: PMC7227725.

7 Genomics and Proteomics

Genomics

Genomics, a branch of molecular biology, studies the complete DNA set (genome) of an organism, including its structure, function, evolution, mapping, and editing. Coined in 1986 by geneticist Thomas Roderick, the term genomics distinguishes itself from conventional genetics, which focuses on individual genes. Genomics takes a broader approach, aiming to characterize and quantify all of an organism's genes, interactions, and their overall impact on the organism. Genes can direct protein production which, in turn, form body structures and control cellular processes, and genomics uses high-throughput DNA sequencing and bioinformatics to sequence, analyse, and understand the function and structure of entire genomes, revolutionizing biology and medicine.

Types of Genomics

Functional Genomics

Functional genomics aims to decipher the roles of genes and their interactions within biological systems. This field investigates gene expression, regulation, and interplay, seeking to identify all gene functions within a genome, map expression patterns across various tissues and conditions, and unravel complex gene interaction networks. This is achieved through various techniques, including transcriptomics, proteomics, gene knockout/ Knockdown studies, epigenomics, etc. By employing these approaches, functional genomics explores gene regulation, elucidates disease mechanisms, and contributes to the development of targeted therapies.

Structural Genomics

Structural genomics comprehensively analyses entire genomes by mapping, sequencing, and determining the 3D structures of encoded proteins. This involves creating high-resolution genetic maps, identifying all genes, and understanding DNA's structural organization, thus providing a basis for understanding gene and protein function. Techniques like X-ray crystallography (XRC), nuclear magnetic resonance (NMR), and genome sequencing are employed. More broadly, structural genomics aims to define the complete genomic structure, including the physical arrangement of genes and other elements. This comprehensive approach facilitates identifying disease-associated genetic variants, understanding chromosomal abnormalities, and developing genome editing techniques.

Comparative Genomics

Comparative genomics analyses the genomes of different species to understand evolutionary relationships, identify conserved genes and regions, and explore the genetic basis of phenotypic differences. Its goals include tracing evolutionary history, pinpointing genes responsible for unique traits, and understanding genome evolution. Key techniques include genome alignment, phylogenetic analysis, and comparative mapping. This field also focuses on identifying orthologs (conserved genes across species) and paralogs (genes evolved within a species), studying conserved genetic elements for insights into essential biological functions, and investigating genome evolution through processes like gene duplication, mutation, and horizontal gene transfer. Ultimately, comparative genomics helps identify disease-related genes, enhances our understanding of evolutionary biology, and aids in developing model organisms for research.

Translational Genomics

Translational genomics bridges the gap between genomic research and clinical practice to enhance patient care. It encompasses using genomic data to personalize treatments, developing diagnostic tests for disease-associated mutations, and studying how genetic variations impact drug response (pharmacogenomics). It studies how an individual's genes influence their response to medications, aiming to personalize drug therapies based on genetic profiles. This field is crucial for diagnosing genetic disorders, predicting disease risk, and developing targeted therapies.

Metagenomics

Metagenomics analyses genetic material recovered directly from environmental samples, enabling the study of microbial communities without needing to isolate or culture individual species. This approach allows researchers to characterize microbial diversity across various environments, identify novel genes and functions from uncultured microorganisms, and understand the role of microbial communities in ecosystem processes. This has applications in diverse fields such as environmental monitoring, agriculture, and human microbiome research.

Bioinformatics in Genomics

Genomics Database

GenBank

GenBank (https://www.ncbi.nlm.nih.gov/genbank/), a comprehensive and freely accessible repository of nucleotide sequences and their corresponding protein translations is a cornerstone of genomics research and is maintained by the National Centre for Biotechnology Information (NCBI), it is part of the International Nucleotide Sequence Database Collaboration, working in conjunction with the DNA Databank of Japan (DDBJ) and the European Nucleotide Archive (ENA) to ensure global data consistency and coverage through daily data exchange.

GenBank's Key Features

- **Annotated Sequences:** Providing rich metadata such as gene locations, coding regions, and protein products.
- **Comprehensive Data:** Housing a vast collection of sequences, from genomic DNA and RNA to expressed sequence tags (ESTs), representing diverse organisms.
- **Public Access:** Offering free access to researchers worldwide, fostering collaboration and data sharing.

- **Regular Updates:** Undergoing bi-monthly updates with new submissions to maintain a current and comprehensive database.
- **Powerful Search and Retrieval:** Enabling users to search by various criteria (accession numbers, keywords, organism names) and utilizing tools like BLAST for sequence similarity searches.
- **Data Annotation:** Providing detailed annotations for each sequence entry, including source organism, gene function, and related publications.
- International Collaboration: Partnering with ENA and DDBJ for global data coverage and consistency.

Applications of GenBank in Genomics

- **Genome Annotation:** Assisting in identifying genes and functional elements in newly sequenced genomes.
- **Comparative Genomics:** Facilitating comparative studies across species to understand evolutionary relationships.
- **Variant Analysis:** Enabling the identification of genetic variants and their implications in health and disease.
- **Metagenomics:** Supporting studies of microbial diversity by providing sequences from various environments.
- **Genome Assembly and Annotation:** Serving as a crucial resource for genome sequencing projects, aiding in comparison, identification, and assembly.
- **Functional Genomics:** Providing information on gene expression, protein structure, and protein-protein interactions to understand gene roles.
- **Medical and Pharmaceutical Research:** Playing a vital role in identifying disease-causing genes, developing diagnostics, discovering drug targets, and understanding drug response for personalized medicine.

EMBL-Bank

EMBL-Bank (https://www.ebi.ac.uk/), also known as the EMBL Nucleotide Sequence Database, is a key resource for genomics research, providing a comprehensive collection of nucleotide sequences. Maintained by the European Molecular Biology Laboratory's European Bioinformatics Institute (EMBL-EBI), it is part of the International Nucleotide Sequence Database Collaboration, working with GenBank (NCBI) and the DNA Databank of Japan (DDBJ) to ensure a globally synchronized repository of genetic data through daily data exchange.

Key Features of EMBL-Bank

- **Annotated Sequences:** EMBL-Bank provides annotated sequences, including information about gene locations, coding regions, and protein products.
- **Public Access:** The database is freely accessible to researchers worldwide, promoting collaboration and data sharing.

- **Regular Updates:** EMBL-Bank is updated regularly with new submissions from researchers, ensuring the database remains current and comprehensive.
- **Search and Retrieval:** Users can search for sequences using various criteria, such as accession numbers, keywords, and organism names, and tools like BLAST (Basic Local Alignment Search Tool) facilitate sequence similarity searches.

Applications of EMBL-Bank in Genomics

- **Genome Annotation:** Providing raw sequence data for assembling and annotating new genomes, and identifying genes and functional elements.
- **Comparative Genomics:** Enabling cross-species comparisons to study evolutionary relationships and functional similarities.
- **Variant Analysis:** Supporting the identification and study of genetic variants and their implications for health and disease.
- **Functional Genomics:** Assisting in the identification of genes, regulatory regions, and other functional genomic elements.
- **Metagenomics:** Providing sequences from diverse environments to aid in understanding microbial diversity.
- **Medical Genomics:** Supporting research into genetic diseases, personalized medicine, and drug discovery.

DDBJ

The DNA Data Bank of Japan (DDBJ) (https://www.ddbj.nig.ac.jp/index-e.html), located at the National Institute of Genetics (NIG) in Shizuoka, Japan, is a crucial resource for genomics research. As a founding member of the International Nucleotide Sequence Database Collaboration (INSDC), alongside GenBank (NCBI, USA) and the European Nucleotide Archive (ENA, EBI, UK), DDBJ ensures global data accessibility by exchanging data daily with its partners. This collaboration synchronizes data within 24 hours, making submissions to any of the three databases available worldwide.

Key Features of DDBJ

- **Annotated Sequences:** Providing information on gene locations, coding regions, and protein products.

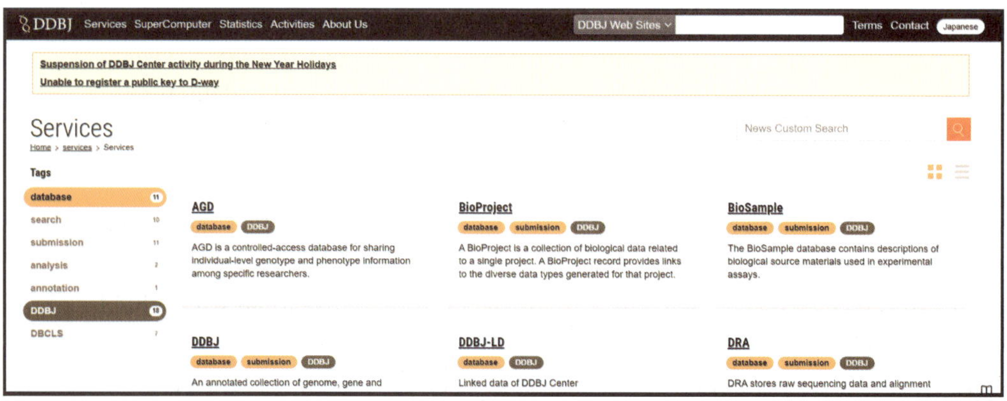

- **Data Collection and Archiving:** Collecting and archiving nucleotide sequences, primarily from Japan but also internationally, and assigning unique accession numbers for easy retrieval.
- **Data Dissemination:** Offering free access to its data via online tools and services for various research purposes, such as comparative genomics, evolutionary studies, and drug discovery.
- **Value-Added Services:** Including the DDBJ Read Archive (DRA) for raw sequencing data, the DDBJ Bio Sample and BioProject Databases for sample and project information, and the Gene Expression Archive (GEA).
- **Regular Updates and Search Functionality:** Regularly updating with new submissions and offering robust search capabilities using various criteria and tools like BLAST.

Applications of DDBJ in Genomics

- **Genome Annotation:** Assisting in identifying genes and functional elements in newly sequenced genomes.
- **Comparative Genomics:** Facilitating evolutionary studies through sequence comparisons across species.
- **Variant Analysis:** Supporting the identification and study of genetic variants.
- **Metagenomics:** Providing sequences from diverse environments to understand microbial diversity.
- **Promoting Collaboration and Data Sharing:** Fostering global research collaboration through open access.
- **Enabling Large-Scale Genomic Studies:** Supporting comprehensive analyses like GWAS and metagenomics research.
- **Supporting Biodiversity Research:** Housing diverse sequence data for understanding biodiversity and evolutionary relationships.
- **Driving Technological Advancements:** Contributing to the development of new bioinformatics tools and algorithms.

Ensembl

A cornerstone of modern genomics, is a comprehensive genome browser and annotation resource developed by the European Molecular Biology Laboratory (EMBL) and the European Bioinformatics Institute (EBI). Launched in 1999 as a collaboration with the Wellcome Trust Sanger Institute, Ensembl offers an integrated genomic database, primarily focused on vertebrates and human-related species.

Key Features of Ensembl

Genome Browser: An intuitive interface for visualizing and navigating entire genomes, displaying gene annotations, transcripts, variations, regulatory elements, and comparative genomics data.

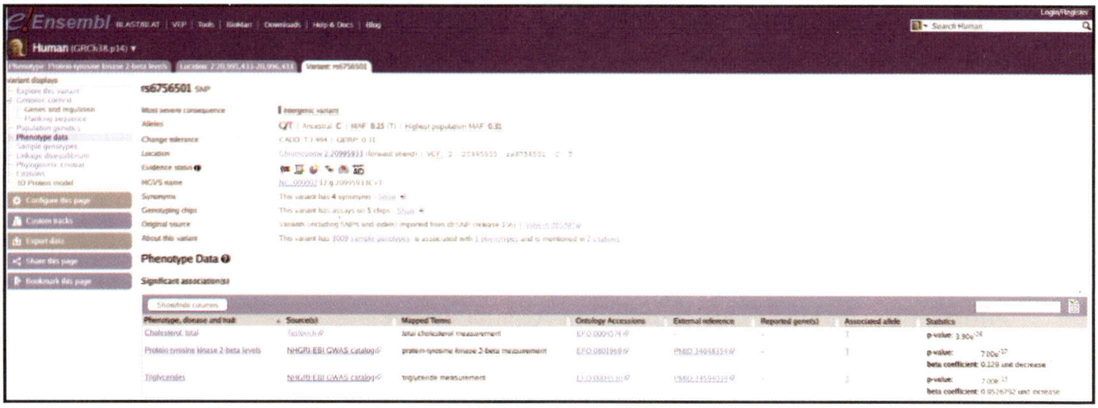

Figure 2: Result page of Ensembl

- **Annotation:** High-quality, automated gene annotation, providing evidence-based gene models and functional information, including gene structure, function, and expression.
- **Comparative Genomics:** Pre-computed genome alignments across multiple species, facilitating evolutionary studies, identifying conserved regions, and understanding functional implications of genomic differences.
- **Variation Data:** Integrated data on genetic variations (SNPs, indels, structural variants) crucial for understanding the genetic basis of diseases and other traits.
- **Regulatory Genomics:** Annotations of regulatory elements (promoters, enhancers, transcription factor binding sites) to understand gene expression regulation and its phenotypic effects.
- **Data Integration:** Aggregation of data from various public databases and research publications, providing a comprehensive and up-to-date view of genomic information.
- **Tools and Resources:** A suite of tools, including BLAST, BLAT, BioMart, and the Variant Effect Predictor (VEP), to support genomic analysis.

Applications of Ensembl in genomics

- **Gene Discovery and Annotation:** Providing a comprehensive view of gene structures and functions.
- **Genome Organization and Function Studies:** Enabling research on the structure, function, and evolution of genomes.
- **Comparative Genomics Research:** Facilitating evolutionary studies and identification of conserved elements.
- **Disease Gene and Variant Identification:** Assisting in identifying genetic variants associated with diseases.
- **Gene Regulation Studies:** Helping understand how gene expression is controlled and its role in development and disease.
- **Variant Analysis:** Analysing genetic variants and their implications for health and diseases.

- **Regulatory Element Studies:** Investigating non-coding regulatory elements and their roles in gene regulation.
- **Functional Genomics:** Providing resources for functional genomics studies, including gene expression analysis and regulatory network construction.

Genomic Software /Tools

Blast

BLAST (Basic Local Alignment Search Tool) (https://blast.ncbi.nlm.nih.gov/Blast.cgi), is a widely used bioinformatics tool for comparing biological sequences like DNA, RNA (nucleotides), and proteins (amino acids). Developed in 1990 by Stephen Altschul and colleagues, it's essential for genomics research. BLAST searches compare a query sequence against a sequence database to find similar sequences above a given threshold.

Key Features of BLAST

- **Sequence Alignment:** BLAST identifies regions of local similarity and calculates their statistical significance.
- **Local Alignment:** It focuses on finding similar regions (local alignments) rather than aligning entire sequences, which is useful for finding conserved domains or motifs.
- **Heuristic Algorithm:** BLAST uses a fast heuristic approach, which, while not guaranteeing the absolute best alignment, provides good results quickly, crucial for large databases.
- **Types of BLAST:** Different BLAST variations exist for specific comparisons:
    - **BLASTN:** Compares nucleotide sequences to nucleotide sequence databases.
    - **BLASTP:** Compares protein sequences to protein sequence databases.
    - **BLASTX:** Compares a nucleotide query sequence translated in all six reading frames to a protein sequence database.
    - **TBLASTN:** Compares a protein query sequence to a nucleotide sequence database translated in all six reading frames.
    - **TBLASTX:** Compares the six-frame translations of a nucleotide query sequence against the six-frame translations of a nucleotide sequence database
- **Statistical Significance:** BLAST uses a heuristic algorithm for efficient searching and reports the E-value (expect value) to assess the statistical significance of matches.

Applications of in Genomics

- **Gene Identification:** Identifying homologous genes and proteins, aiding gene annotation and discovery.
- **Functional Annotation:** Inferring the function of new genes or proteins by comparing them to known sequences.
- **Evolutionary Studies:** Studying evolutionary relationships by identifying conserved sequences across species.

- **Identifying Conserved Domains:** Finding conserved protein domains to understand protein function and evolution.
- **Genome Assembly and Annotation:** Assisting in assembling and annotating new genomes.
- **Pathogen Detection:** Identifying pathogens by comparing sequences to known pathogen databases.
- **Comparative Genomics:** Comparing genomes of different organisms.
- **Drug Discovery:** Identifying potential drug targets by finding similarities between disease-related proteins and known drug targets.

SAM Tools

SAM tools (https://www.htslib.org/), developed by Heng Li, is a suite of command-line utilities essential for processing short DNA sequence read alignments. It works with SAM (Sequence Alignment/Map), BAM (Binary Alignment/Map), and CRAM formats, which are standard in genomics research for storing and managing aligned reads. The SAM format, in particular, is a versatile, compact, and efficient format suitable for various sequencing platforms and read lengths.

Key Features of SAM tools

- **Sorting and Indexing:** Efficiently sorts and indexes large BAM files for rapid access to specific genomic regions.
- **Variant Calling:** Facilitates the identification of genetic variations using tools like bcftools, and identifying variations in the sequenced DNA compared to the reference sequence.
- **Alignment Viewing and Manipulation:** Enables viewing, filtering, and extracting specific reads from alignments.
- **Format Conversion:** Converts between SAM, BAM, and other formats.
- **Sorting and Merging:** Sorts alignments by coordinate or read name and merges multiple alignment files.
- **Data Extraction:** Extracts specific genomic regions from alignment files.
- **Compression:** Supports space-saving compressed formats like CRAM.

Applications of SAM Tools in Genomics

- **Genome Assembly and Resequencing:** Used for assembling genomes from short reads and identifying variations compared to a reference.
- **Variant Analysis:** Aids in identifying and analysing genetic variants for studies on genetic diseases and population genetics.

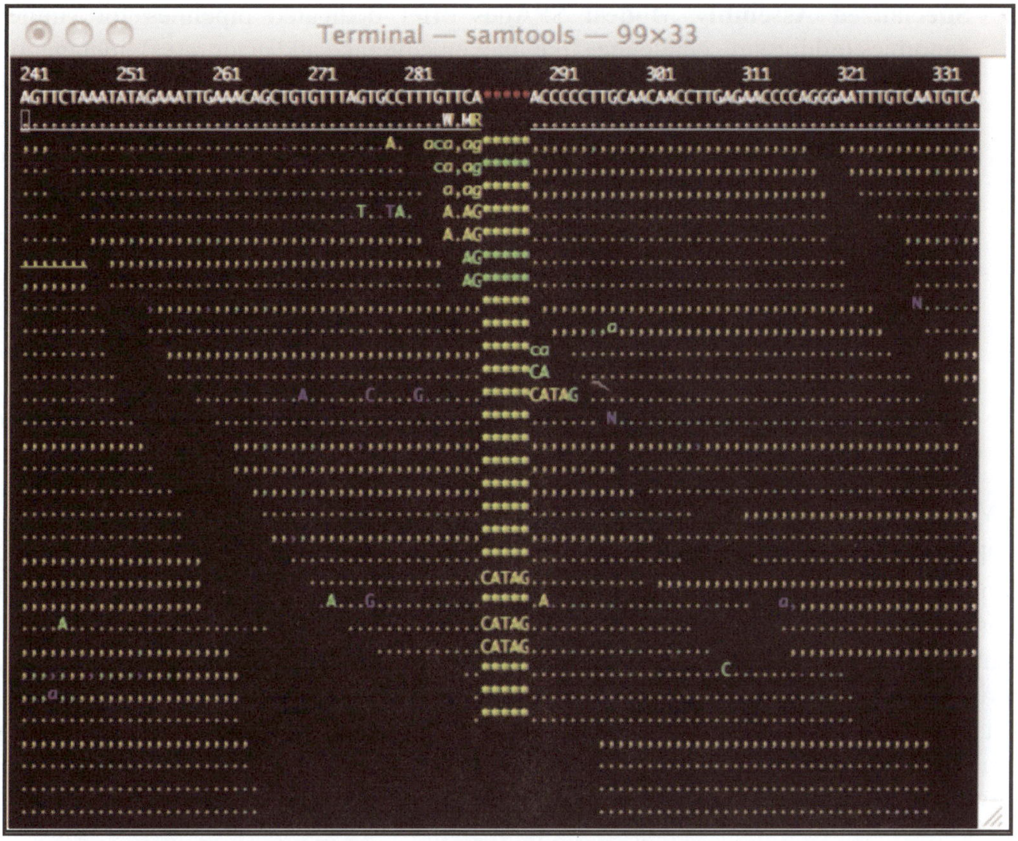

Figure 3: Variant calling in SAMtools

- **Gene Expression Studies (including RNA-Seq):** Supports analysis of gene expression levels and transcript isoforms by filtering and extracting relevant reads.
- **Comparative Genomics:** Enables comparison of genomic regions across different species.
- **Quality Control:** Provides tools for assessing and improving the quality of sequencing data.
- **ChIP-Seq:** Used for mapping protein-DNA interactions.

SPAdes

SPAdes (St. Petersburg genome assembler) (https://ablab.github.io/spades/index.html), is a versatile and widely used de novo genome assembly toolkit in genomics, adept at handling diverse sequencing data and applications. SPAdes excel at assembling genomes from single-cell, metagenomic, and standard bacterial datasets, and has since been further developed for broader applications.

Key Features of SPAdes

- **Broad Data Compatibility:** It supports various sequencing technologies, including Illumina, Ion Torrent, PacBio, and Oxford Nanopore, and facilitates hybrid assemblies combining short and long reads.

- **Specialized Assembly Modes:** SPAdes offer dedicated pipelines for single-cell genomics (addressing challenges like uneven coverage and error rates), metagenomics (reconstructing genomes from complex microbial communities), plasmid discovery (identifying and assembling plasmids), and RNA viral genomics (assembling viral genomes from transcriptomic data). It also excels at standard bacterial genome assembly.
- **Robust Assembly Approach:** SPAdes employ a de Bruijn graph-based approach, effectively handling uneven sequencing depths and complex genomic regions, resulting in accurate and high-quality assemblies.
- **Versatility, Accuracy, and Efficiency:** These include single-cell and metagenomic assembly pipelines, hybrid assembly support, and specialized modes for plasmid and RNA viral assembly. It is also known for its accuracy, efficiency, and relatively user-friendly command-line interface.

PhyML

The PhyML (Phylogenetic Maximum Likelihood) (http://www.atgc-montpellier.fr/phyml/paper.php), is a widely used genomics software package for inferring evolutionary relationships by constructing phylogenetic trees from molecular sequence data (DNA or protein). It employs the statistically robust maximum likelihood (ML) method, evaluating the probability of observed sequence data under various tree topologies and evolutionary models. In essence, PhyML is a powerful tool for reconstructing evolutionary histories from molecular data, with applications ranging from basic evolutionary research to practical applications in medicine and conservation.

Key Features of PhyML

- **Maximum Likelihood Inference:** Uses ML to estimate phylogenetic trees, providing a probabilistic framework for tree inference.
- **Substitution Models:** Supports a wide range of substitution models to accommodate different evolutionary rates.
- **Tree Search Algorithms:** Employs efficient algorithms like Nearest Neighbor Interchange (NNI), Tree Bisection and Reconnection (TBR), and Subtree Pruning and Regrafting (SPR) to explore possible tree topologies.
- **Evolutionary Models:** Supports diverse evolutionary models to account for varying substitution rates among nucleotides or amino acids.
- **Statistical Support:** Offers bootstrapping and approximate likelihood-ratio tests (aLRT) to assess the statistical confidence in inferred tree topologies and branch support.

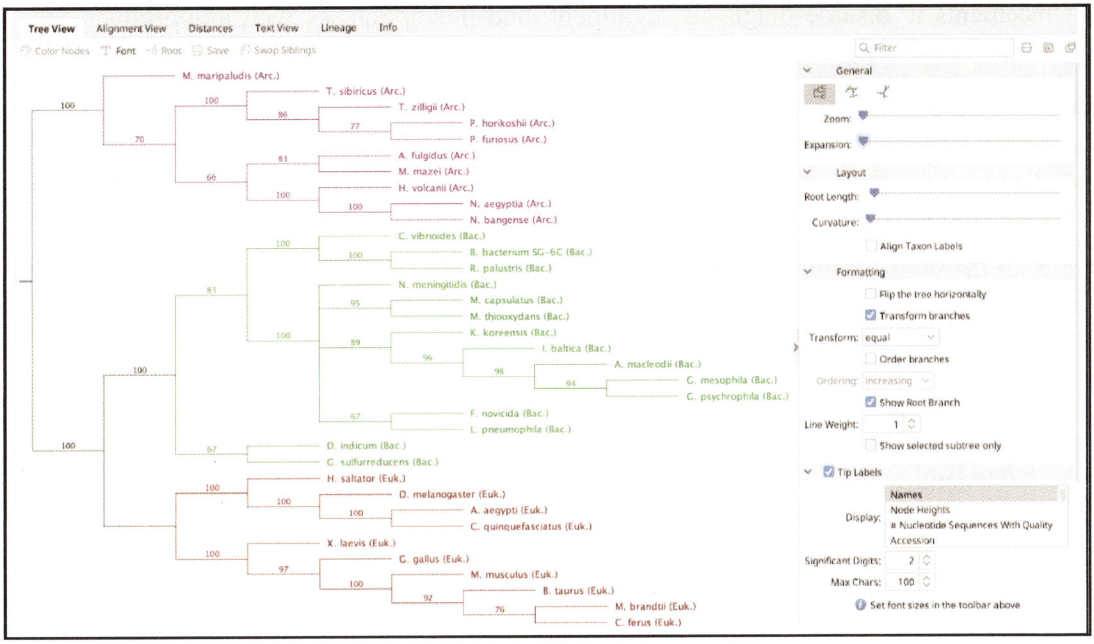

Figure 4: Tree view from PhyML

- **User Interface:** Provides both a graphical user interface (GUI) and command-line options.
- **Parallel Computing:** Supports parallel processing for faster analysis of large datasets.

Applications of PhyML in Genomics

- **Phylogenetic Tree Construction:** Reconstructing evolutionary relationships between species or genes.
- **Evolutionary Biology:** Understanding species origins, adaptation, and diversification.
- **Infectious Disease Epidemiology:** Tracking pathogen transmission, identifying outbreak sources, and monitoring drug resistance.
- **Conservation Biology:** Assessing genetic diversity and informing conservation strategies.
- **Comparative Genomics:** Studying gene and genome evolution across species, identifying conserved regions and functional elements, and understanding the genetic basis of phenotypic diversity.
- **Population Genetics:** Analysing genetic variation and evolutionary dynamics within and between populations.
- **Functional Genomics:** Studying gene function and regulation through evolutionary analysis.

Application Of Genomics

Genomics has profoundly transformed diverse scientific and technological fields. By deciphering the intricate genetic blueprint of living beings, genomics has paved the way for groundbreaking advancements in disease diagnosis, treatment, and prevention, as well as improvements in agriculture, forensics, environmental studies, and social sciences.

Medical Applications

Genomics is at the forefront of personalized medicine, enabling tailored treatment strategies based on an individual's unique genetic makeup. This approach enhances the effectiveness and safety of medical interventions. Key applications include:

- **Precision Medicine and Pharmacogenomics:** By analysing an individual's genetic profile, pharmacogenomics optimizes drug therapies, minimizing adverse drug reactions and maximizing efficacy. For example, testing for CYP2C19 gene variants informs optimal dosing of the antiplatelet drug clopidogrel.
- **Diagnosis and Risk Prediction:** Genomics plays a crucial role in diagnosing genetic disorders, predicting disease susceptibility, and guiding reproductive decisions. Genetic testing can identify mutations associated with hereditary cancers, such as BRCA1/BRCA2 for breast and ovarian cancer.
- **Gene Therapy:** Genomics is fundamental to the development of gene therapies, which aim to correct genetic defects by introducing functional genes into patient cells. An example is the treatment of spinal muscular atrophy (SMA) using gene therapy (onasemnogene abeparvovec).
- **Cancer Genomics:** Identifying the genetic mutations driving cancer development allows for targeted therapies and personalized treatment plans. For example, targeted therapies for non-small cell lung cancer based on specific genetic mutations and the use of imatinib for chronic myeloid leukemia (CML) based on the presence of the BCR-ABL fusion gene.

Biotechnology

Genomics has significantly impacted biotechnology, particularly in agriculture and industrial applications:

- **Agriculture:** Genomics facilitates the development of genetically modified crops with improved yield, nutritional content, pest resistance (e.g., Bt cotton), and tolerance to environmental stressors (e.g., drought-resistant maize varieties). It also aids in identifying animals with superior genetic traits for meat production, milk yield, and disease resistance, informing breeding decisions and enhancing livestock productivity (e.g., genomic selection in dairy cattle).
- **Industrial Biotechnology:** Genomics supports the development of microorganisms for producing biofuels (e.g., genetically engineered yeast for bioethanol production), pharmaceuticals, and other bioproducts.

Environmental Studies

Genomics provides valuable tools for studying and protecting the environment:

- **Conservation Biology:** Genomics helps understand the genetic diversity and evolutionary history of endangered species, informing conservation strategies. For instance, genomic studies have assessed the genetic health of the endangered Florida panther population.

- **Microbial Ecology:** Genomics, particularly metagenomics, is used to study microbial communities in various environments, providing insights into ecosystem functions and responses to environmental changes. This includes studying microbial diversity in soil ecosystems and its impact on nutrient cycling.

Forensic Science

Genomics forms the basis of DNA fingerprinting, a powerful tool for individual identification:

- **DNA Profiling:** DNA profiling, based on unique DNA profiles, is invaluable in criminal investigations, paternity testing, and identifying victims. It helps link suspects to crime scenes and has even been used to exonerate wrongfully convicted individuals.

Challenges in Genomics

Genomics research, while rapidly advancing, faces several key challenges that impede its progress. These challenges span various areas, including samples, data, ethics, technology, collaboration, diversity, and clinical translation.

- **Sample Quality, Quantity, and Diversity:** High-quality DNA samples are essential for accurate results. Degradation of DNA can lead to errors. Also, statistically significant results often require large sample sizes, which can be difficult to obtain, especially for rare diseases. And, a significant bias exists towards studies focusing on European ancestry populations. This lack of diversity limits the generalizability of findings to other ethnic groups and can exacerbate health disparities.
- **Data Analysis and Interpretation:** Analysing the billions of base pairs in a single human genome requires substantial computational resources and expertise, efficient storage, management, and analysis of this data are crucial. Further, identifying and interpreting the functional significance of genetic variations is a major challenge, many variants have unknown significance, hindering translation to clinical practice. Also, analyzing genomic data requires sophisticated bioinformatics tools and skilled personnel. Along with that, integrating genomic data with other data types (electronic health records, imaging data) is crucial but challenging due to data heterogeneity and lack of standardization.
- **Ethical and Legal Issues:** Protecting the privacy of individuals' highly sensitive genetic information is paramount, especially with increasing direct-to-consumer testing and large genomic databases. Also, obtaining truly informed consent for genomic research is complex due to the potentially unforeseen implications of the data and its future uses. There is a risk of genetic discrimination in areas like employment and insurance based on genetic predispositions.
- **Technological and Resource Limitations:** While sequencing costs have decreased, they remain a barrier for many research studies and clinical applications, especially in resource-limited settings. Not all institutions have access to the latest genomic technologies, limiting the scope of studies. Despite advancements, sequencing technologies can still produce errors, affect analysis accuracy, and determine the functional consequences of identified genetic variations remains a significant challenge.

- **Interdisciplinary Collaboration:** Effective collaboration between different scientific disciplines is essential for integrating and interpreting genomic data.
- **Lack of Diversity in Study Populations:** As mentioned earlier, the overrepresentation of European ancestry populations limits the generalizability of findings and contributes to health disparities.
- **Translating Research into Practice:** Translating genomic research findings into clinical practice requires further research to validate clinical utility and develop usage guidelines.

Overall, while genomics holds tremendous promise for advancing healthcare, addressing these multifaceted challenges is crucial for realizing its full potential and ensuring equitable benefits for all populations.

Future Prospectus of Genomics

The field of genomics is undergoing rapid advancements, promising transformative changes across various sectors. The future of genomics holds immense potential, particularly in healthcare, agriculture, and environmental science.

- **Personalized Medicine:** Genomics will further personalize treatments based on individual genetic profiles, leading to more effective and safer interventions. This includes analyzing a patient's DNA to predict drug response, optimize dosages, and minimize adverse reactions, selecting the most effective drugs and dosages based on individual genetic variations, identifying specific genetic mutations in tumors to enable targeted therapies, and predicting disease risk for early interventions and preventive measures.
- **Disease Prediction and Prevention:** Genomics enables proactive healthcare by identifying individuals at high risk for diseases like cancer, heart disease, and Alzheimer's for early intervention. Also, expanding screening to detect a wider range of genetic disorders for early diagnosis and treatment, and advancing non-invasive prenatal testing (NIPT) to detect a broader range of genetic conditions.
- **Drug Discovery and Development:** Genomics accelerates and improves drug development by uncovering genes and pathways involved in disease development to identify novel drug targets, identifying existing drugs effective against other diseases based on shared genetic pathways, and classifying patients based on genetic profiles to improve trial efficiency and success.
- **Genome Editing:** Technologies like CRISPR-Cas9 will enable precise genome modifications for correcting mutations responsible for genetic disorders like sickle cell anemia and cystic fibrosis.
- **Metagenomics:** Studying genetic material recovered directly from environmental samples will advance our understanding of the human microbiome's role in health and disease for new treatments and dietary recommendations and studying microbial communities for conservation and understanding of ecosystem dynamics.

- **Single-Cell Genomics:** Analyzing the genomes of individual cells will provide insights into understanding how individual cells contribute to health and disease, studying developmental processes and cell differentiation.
- **3D Genomics:** Studying the spatial organization of the genome will reveal insights into gene regulation and chromosomal interactions, and understanding the mechanisms underlying various diseases, including cancer.
- **Agriculture and Food Security:** Genomics will improve the development of crops resistant to pests, diseases, and environmental stressors. Also, identifying animals with desirable traits for improved breeding and productivity, and tracing the source of foodborne outbreaks for rapid identification and control.
- **Environmental Science:** Genomics will contribute to assessing and monitoring biodiversity and informing conservation efforts, detecting pollutants, assessing climate change impact, and developing bioremediation strategies. Further, designing and engineering new biological systems for applications like biofuel production and environmental cleanup.
- **Integration with AI and Big Data:** Combining genomics with AI and big data will enhance the processing and interpretation of large genomic datasets, and improve the accuracy of genetic predictions and personalized therapies.

Case Studies In Genomics

Deciphering the Genetic Underpinnings of a Rare Disease (Sotos Syndrome)

Background: Rare diseases, also known as orphan diseases, are defined as conditions that affect a small percentage of the population. While each individual rare disease is uncommon, collectively they impact a significant number of individuals worldwide. A substantial proportion of rare diseases have a genetic origin, meaning they are caused by alterations in an individual's DNA. This genetic basis makes genomics an exceptionally valuable tool for investigating these conditions, offering the potential to pinpoint the specific genetic mutations responsible.

Case Presentation: This case study involves a young child who presented with a complex set of symptoms. These symptoms included developmental delays, unusual facial features, and heart abnormalities. Despite thorough clinical examination and standard diagnostic tests, the underlying cause of the child's condition remained elusive. Given the complexity and unusual combination of symptoms, clinicians suspected a genetic disorder and opted for whole-exome sequencing (WES).

Genomic Analysis: Whole-exome sequencing (WES) is a genomic technique that focuses on sequencing the protein-coding regions of the genome, known as exons. These exons constitute approximately 1% of the total DNA but contain the vast majority of known disease-causing mutations. In this case, WES analysis uncovered a novel heterozygous mutation in a gene called NSD1. A heterozygous mutation means that the child inherited one mutated copy of the gene and one normal copy. The NSD1 gene is known to play a crucial role in chromatin remodelling and transcriptional regulation during development. Prior research has linked mutations in NSD1 to Sotos syndrome, a rare genetic disorder characterized by excessive growth during childhood, distinctive facial features, and learning disabilities.

Interpretation and Diagnosis: The identification of this novel NSD1 mutation provided a definitive molecular diagnosis for the child's condition. While the child's symptoms did not perfectly align with the classic presentation of Sotos syndrome, the genetic finding offered a clear explanation for their developmental problems. This highlights an important aspect of genomic diagnostics; individuals with mutations in the same gene can exhibit a range of symptoms. The genetic diagnosis also had important implications for genetic counselling. By understanding the genetic basis of the child's condition, the parents could receive accurate information about the recurrence risk for future pregnancies.

Conclusion: This case study powerfully demonstrates the transformative potential of genomics in diagnosing rare diseases, especially those with complex and atypical presentations. By identifying a novel mutation in the NSD1 gene, clinicians were able to provide a precise molecular diagnosis for a child with developmental delays and unusual features, where traditional diagnostic methods had failed. This diagnosis not only elucidated the underlying cause of the child's condition but also provided crucial information for genetic counselling and family planning, empowering the family with knowledge and enabling informed decision-making.

This case study underscores the profound impact of genomics in modern medicine, particularly in the diagnosis and management of rare diseases. As technology continues to advance and our understanding of the genome deepens, genomics will undoubtedly play an increasingly vital role in improving human health.

Protomics

Proteins are biomolecules built from amino acids and perform numerous critical functions, such as providing structural support, driving metabolism, facilitating transport, enabling immunity, mediating signalling, and regulating various processes. Proteomics involves the large-scale study of proteins, examining their structures, functions, and interactions. This complex and rapidly developing field are crucial to functional genomics and was coined in 1997 by Marc Wilkins as an analogy to genomics. Proteomics seeks to identify and quantify all proteins within a biological system; from cells to entire organisms. A key aspect of proteomics is the study of post-translational modifications (PTMs) like phosphorylation, glycosylation, and ubiquitination, which significantly impact protein activity, location, and interactions. Furthermore, recognizing that proteins function within intricate networks, proteomics investigates how they interact to form complexes and carry out essential cellular processes. However, it is much more complicated than genomics because the protein expression is altered according to time and environmental conditions.

Types of Proteomics

Proteomics can be categorized into three main types; expression proteomics, functional proteomics, and structural proteomics, each focusing on different aspects of protein study.

Expression Proteomics

This type of proteomics is a unique way to study quantitative and qualitative protein expression and it aims to identify the difference in protein expression between two samples or conditions (Banks et al., 2000). Expression proteomics assays are commonly used to investigate protein expression patterns in different cells. To detect variations in protein levels, a tumour tissue sample is compared to normal tissue samples. It involves techniques such as two-dimensional gel electrophoresis (2-DE), mass spectroscopy (MS), and liquid chromatography-mass spectroscopy (LC-MS). It has a wide range of applications such as in identifying biomarkers for disease diagnosis and prognosis, comparative protein expression studies, or understanding cellular responses to various stimuli (drugs, stress).

Functional Proteomics

This area of proteomics explores how proteins function within cells and the molecular processes they drive such as in PTM; specifically identifying the proteins they interact with. It focuses on understanding how an uncharacterized protein connects with other proteins within a specific complex involved in a particular cellular activity. This analysis can reveal the unknown protein's biological role. It consists identifying protein complexes and their roles in specific biological processes. Techniques such as affinity chromatography coupled with MS, protein microarray, and yeast two-hybrid screening are often used in this approach. By using this, we can map protein interaction networks, understand signal transduction pathways, and identify drug targets and mechanisms of drug action.

Structural Proteomics

This field of proteomics is concerned with determining the three-dimensional structure of proteins and understanding the precise arrangement of atoms within a protein that helps in studying protein-protein interactions and interactions with other biomolecules like DNA or RNA or membranes, cell organelles, and ribosomes in the mixture. An instance of structural proteomics is the investigation of the nuclear pore complex. Techniques such as X-ray crystallography, Cryo-electron microscopy (Cryo-EM), and nuclear magnetic resonance (NMR) spectroscopy are employed in this approach. Understanding protein function at the molecular level, rational medication creation targeting specific protein structures, and studying protein folding and misfolding disorders can be achieved by this type of proteomics.

Bioinformatics in Proteomics

Bioinformatics is crucial to proteomics research because it provides the tools necessary to analyze and interpret the large and complex datasets generated by proteomics technologies. Without sophisticated computational approaches, understanding this data would be impossible. A wide range of bioinformatics databases and software are used in proteomics research.

Proteomics Database

PDB

The Protein Data Bank (PDB) (https://www.rcsb.org/) is a fundamental resource for structural biology and proteomics. This global repository archives experimentally determined 3D structures of large biological molecules like proteins, nucleic acids, and their complexes, which are crucial for understanding protein function, interactions, and biological roles.

Key Features of the PDB

- **Data Deposition and Curation:** PDB provides access to experimentally determined 3D structures obtained through techniques such as X-ray crystallography, NMR spectroscopy, and cryo-electron microscopy (cryo-EM).

Figure 5: PDB Webpage

- **Data Format and Accessibility:** Using the standardized PDB file format ensures data consistency and facilitates analysis. The PDB is freely accessible online through the Worldwide Protein Data Bank (wwPDB) consortium (US, Europe, and Japan).
- **Computed Structure Models (CSMs):** The PDB includes CSMs generated by computational methods like AlphaFold, providing structural information for proteins lacking experimental structures.
- **Data Validation and Annotation:** Expert curators validate and annotate deposited structures before public release, ensuring data quality and accuracy. Annotations include sequence, function, ligand, and experimental details.

Role of the PDB in Proteomics

- **Protein Structure Prediction and Modeling:** The PDB serves as a training and validation resource for computational methods predicting protein structures from sequence data.
- **Understanding Protein Function:** 3D structures reveal crucial insights into protein function by showing active sites, binding pockets, and overall shape, explaining molecular interactions and biological roles.
- **Drug Discovery and Design:** Studying disease-related protein structures in the PDB enables researchers to identify drug targets and design inhibitory molecules.
- **Comparative Proteomics:** Comparing protein structures across organisms or conditions reveals evolutionary relationships and functional differences.
- **Structural Genomics:** The PDB is the central repository for structures determined by structural genomics initiatives, which aim to determine the structures of all proteins encoded by a genome.

UniProt

UniProt (Universal Protein Resource) (https://www.uniprot.org/) is a freely accessible, high-quality, and comprehensive source of protein sequence and functional data. As a central protein data hub, it integrates various databases to offer a unified perspective on protein information. This extensive resource encompasses protein sequences, functions, classifications, and cross-references, utilizing expert curation and computational analysis to ensure data accuracy and reliability. UniProt offers data in multiple formats (FASTA, XML, RDF) for easy integration with other bioinformatics tools.

Key Components of UniProt

- UniProt Knowledgebase (UniProtKB): The core of UniProt, this database offers detailed protein information and is divided into:

  1.UniProtKB/Swiss-Prot: A manually curated, high-quality, non-redundant database rich in functional information.

  2.UniProtKB/TrEMBL: A computationally annotated database containing translated coding sequences from public nucleotide databases.

- **UniProt Reference Clusters (UniRef):** These provide clustered protein sequences at 100%, 90%, and 50% sequence identity, reducing redundancy for large-scale analyses.
- **UniProt Archive (UniParc):** A comprehensive repository of all public protein sequences, ensuring no redundancy across different databases.

UniProt's Role in Proteomics

- **Protein Identification:** In mass spectrometry, UniProtKB is the primary database used to compare peptide mass fingerprints or fragment ion spectra for protein identification, offering comprehensive sequence and post-translational modification (PTM) data.
- **Protein Characterization:** UniProt provides extensive functional annotation, including protein names, functions, domains, subcellular localization, PTMs, and interactions, which is essential for understanding protein roles in biological processes.
- **Data Analysis and Interpretation:** UniProt aids in interpreting proteomics data by providing context and background information, facilitating pathway and network analysis, and identifying potential drug targets or biomarkers.
- **PTM Analysis:** UniProt's PTM information is crucial for identifying and characterizing PTMs from proteomics data, as these modifications are key regulators of protein function.

Pfam

Pfam (http://pfam-legacy.xfam.org/) is a widely-used database in proteomics that provides a comprehensive collection of protein families. Each family in Pfam is represented by multiple sequence alignments and hidden Markov models (HMMs). These HMMs are powerful tools that enable the identification of protein domains which are the functional and structural units within proteins.

Key Features of Pfam

- **Domain-centric:** Pfam focuses on classifying proteins based on their constituent domains, offering a modular understanding of protein structure and function.

- **Family and Clan Classification:** It organizes protein families and further groups them into clans, reflecting evolutionary relationships.
- **Data Integration:** Pfam seamlessly integrates with other key databases like UniProt and InterPro, providing a rich source of protein information.
- **User-friendly Tools:** Tools like ScanProsite facilitate easy searching and analysis of protein sequences against Pfam signatures.

Roles of Pfam in Proteomics

- **Function Prediction:** Accurately predicting the function of uncharacterized proteins.
- **Protein Annotation:** Providing detailed annotations of protein sequences, revealing their potential roles and interactions.
- **Evolutionary Analysis:** Understanding the evolutionary relationships between protein families.
- **Homology Detection:** Identifying proteins with similar evolutionary origins.
- **Drug Discovery:** Identifying potential drug targets and designing effective drug molecules.
- **Structural Biology:** Understanding the structural features essential for protein function.
- **Genome Annotation:** Assisting in the annotation of genomes by identifying protein-coding regions and predicting their functions.

KEGG

The Kyoto Encyclopedia of Genes and Genomes (KEGG) (https://www.genome.jp/kegg/) is a comprehensive resource integrating genomic, chemical, and functional information, widely used in various biological fields, including proteomics. It offers valuable tools for analyzing and interpreting large-scale protein datasets.

Key Features of KEGG Databases

- **KEGG PATHWAY:** This database provides manually curated pathway maps illustrating molecular interactions and reaction networks (metabolic and regulatory pathways), enabling visualization and analysis of complex interactions between proteins, genes, and metabolites.
- **KEGG ORTHOLOGY (KO):** KEGG assigns KO identifiers to group genes and proteins into orthologous groups based on functional similarity, facilitating cross-species comparisons and protein functional annotation.
- **KEGG GENES:** This database contains genes and proteins from complete genomes of cellular organisms and viruses, providing annotations based on sequence similarity and functional characterization
- **KEGG COMPOUNDS:** This section details small molecules (metabolites, drugs, and chemical compounds), useful for studying metabolic pathways and drug interactions

- **KEGG DISEASE:** This database links diseases to KEGG pathways, aiding in understanding the molecular basis of diseases and identifying potential therapeutic targets.

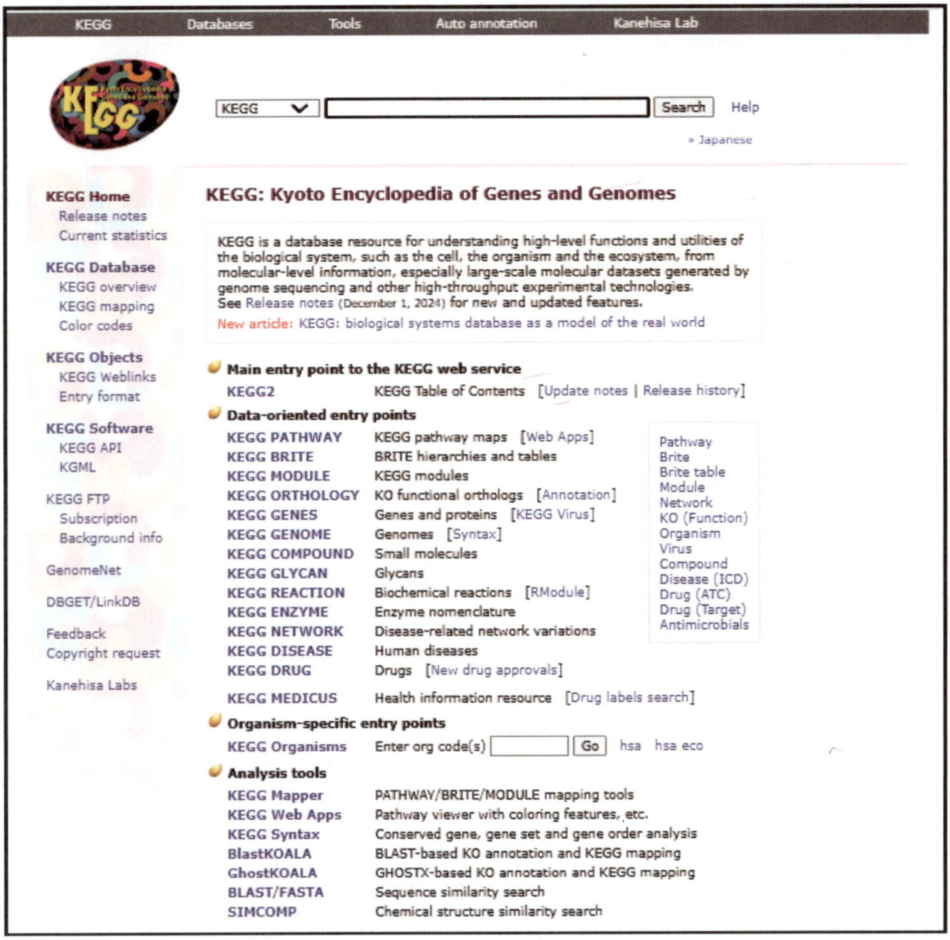

Figure 6: Homepage of KEGG

Role of KEGG in Proteomics

- **Pathway Mapping and Enrichment Analysis:** KEGG's strength lies in its pathway database, offering graphical representations of biological pathways (metabolic, signaling, and disease-related). By mapping differentially expressed proteins from proteomics studies onto KEGG pathways, researchers can identify enriched pathways, revealing significantly affected biological processes and providing functional context for proteomic changes.
- **Functional Annotation:** KEGG provides functional annotations for proteins, including pathway involvement, molecular functions, and associated diseases, crucial for interpreting proteomics data and generating hypotheses about protein roles.
- **Protein-Protein Interaction Networks:** KEGG also contains protein-protein interaction information, enabling the construction of interaction networks to analyze

relationships between proteins identified in proteomics experiments, providing insights into regulatory mechanisms and functional modules.

- **BRITE Database:** The KEGG BRITE database offers hierarchical classifications of biological entities, including proteins, allowing categorization based on function, family, or other properties, aiding in proteomics data interpretation.

Proteomics Software/ Tool

CLUSTAL Omega

CLUSTAL Omega (https://www.ebi.ac.uk/jdispatcher/msa/clustalo) is a widely-used bioinformatics tool that excels at aligning multiple protein or nucleotide sequences to pinpoint areas of similarity across sequences, offering insights into crucial functional or structural elements. This tool determines how proteins have evolved over time, shedding light on their origins and diversification, and gaining insights into the likely function of a protein based on its similarities to proteins with known functions.

Key Features of CLUSTAL Omega

- **Progressive Alignment:** CLUSTAL Omega utilizes a step-by-step approach, first aligning closely related sequences and gradually incorporating more distant ones. This strategy, guided by a tree-like structure representing sequence relationships, ensures efficient and accurate alignment.
- **Versatile Input/Output:** The tool supports a wide range of input formats (FASTA, NBRF/PIR, Clustal) and generates output in various formats (Clustal, Phylip, GCG, NEXUS) suitable for different downstream analyses.
- **Customization Options:** Users can fine-tune the alignment process by adjusting parameters such as gap penalties, scoring matrices, and weighting schemes to optimize results for specific datasets.
- **User-Friendly Interface:** CLUSTAL W is accessible to researchers of all levels, offering both command-line and user-friendly graphical interfaces (like ClustalX).

Roles of CLUTAL Omega in Proteomics

- **Homology Detection:** Uncover evolutionary relationships between proteins across different species, providing valuable insights into functional conservation.
- **Phylogenetic Analysis:** Construct evolutionary trees based on the aligned sequences, revealing the evolutionary history and relationships among proteins.

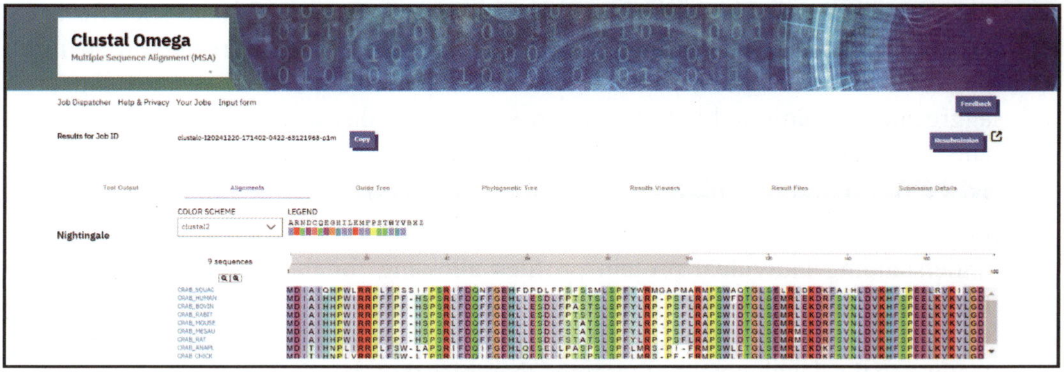

Figure 7: Results of CLUSTAL Omega

- **Functional Site Prediction:** Identify functionally important sites within proteins, such as active sites, binding sites, and sites for post-translational modifications, by focusing on conserved regions.
- **Protein Structure Prediction:** Enhance the accuracy of protein structure prediction by incorporating information from the alignment of homologous sequences.
- **Drug Design:** Aid in the discovery and development of new drugs by facilitating the analysis of protein-ligand interactions and the identification of potential drug targets.

Python libraries

Python is a powerful programming language that is widely used in data analysis, scientific computing, and many other fields. Several libraries enhance Python's capabilities, making it easier to handle complex data and perform advanced computations. The three of the most popular libraries: NumPy, Pandas, and SciPy.

1. NumPy (Numerical Python) (https://numpy.org/) is a fundamental package for scientific computing in Python. It provides support for large, multi-dimensional arrays and matrices, along with a collection of mathematical functions to operate on these arrays.
2. Pandas (https://pandas.pydata.org/) is a library for data manipulation and analysis. It provides data structures like DataFrame and Series, which are designed to handle structured data efficiently.
3. SciPy (Scientific Python) (https://docs.scipy.org/doc/scipy/reference/)is a library that builds on NumPy and provides additional tools for scientific and technical computing. It includes modules for optimization, integration, interpolation, linear algebra, signal processing, etc.

Application of Proteomics

Proteomics has emerged as a powerful tool with diverse applications across various fields.

- **Disease Biomarker Discovery:** Proteomics is a powerful tool for uncovering biomarkers and measurable indicators of the biological state of various diseases. By analyzing the protein composition of cells, tissues, and bodily fluids, we can identify unique protein signatures associated with specific diseases like cancer, cardiovascular diseases, and neurodegenerative disorders. For instance, prostate-specific antigen (PSA) for prostate cancer and CA-125 for ovarian cancer were identified through proteomic studies. Also, Proteomics has been instrumental in identifying protein aggregates like amyloid-beta and tau in Alzheimer's disease, leading to new diagnostic and therapeutic strategies

- **Drug Development:** Proteomics significantly contributes to drug development by identifying and validating drug targets, assessing drug efficacy and toxicity, and understanding individual protein profiles allowing for the development of targeted therapies tailored to specific patient characteristics. For example, in breast cancer research, proteomics has identified specific proteins driving cancer progression, leading to the development of targeted therapies that specifically inhibit these proteins.
- **Personalized Medicine:** Proteomics is valuable in personalized medicine, enabling treatment strategies tailored to individual patients. By analyzing individual protein profiles, clinicians can predict individual responses to treatments, and minimize adverse reactions. In cystic fibrosis, for instance, proteomic analyses have identified specific protein mutations that can be targeted with personalized therapies, leading to improved patient outcomes.
- **Functional Genomics:** Proteomics complements genomics by bridging the gap between gene expression and protein function. While genomics studies genes, proteomics investigates the resulting proteins, providing insights by revealing how gene expression translates into functional proteins, the roles of proteins in cellular processes, and the effects of gene mutations on protein function like how mutations alter protein expression, structure, and activity, leading to a better understanding of disease mechanisms.
- **Structural Biology:** Proteomics employs techniques like mass spectrometry and X-ray crystallography to determine the three-dimensional structures of proteins. This structural information is crucial for understanding protein function and designing drugs that modulate protein function. The structural analysis of the HIV-1 protease, for example, has been crucial for developing effective antiretroviral drugs for HIV/AIDS.
- **System Biology:** Proteomics is an integral part of systems biology, which aims to understand biological systems as a whole. By studying protein interactions and networks, we can elucidate complex biological processes, identify key regulatory proteins, and understand cellular responses to stimuli which provides insights into cellular adaptation and disease development. For instance, in studying metabolic pathways, proteomics can identify key regulatory proteins and their interactions, revealing potential therapeutic targets.
- **Agriculture and Food Science:** Proteomics has numerous applications in agriculture and food science, such as improving crop resistance to diseases and environmental stress by identifying proteins involved in plant defense mechanisms, leading to the development of more resilient crops; Assessing food quality and safety by detecting contaminants, allergens, molds, and other factors affecting food quality and safety; Monitoring food spoilage and contamination by analyzing protein changes during food storage and processing can help prevent spoilage and ensure food safety; Improving

crop yields by studying protein networks involved in plant growth and development can lead to strategies for increasing crop productivity. For example, proteomics can detect allergens like peanuts and milk in food products and authenticate food products to ensure quality.

- **Environmental Monitoring:** Proteomics plays a vital role in environmental monitoring by analysing protein responses to pollutants and other environmental factors that can help to assess the health of ecosystems, identifying specific proteins that are affected by pollutants, and providing sensitive indicators of environmental contamination. Proteomics can characterize the protein composition of microbial communities, providing insights into their roles in ecosystems and their responses to environmental changes. For example, studying the effects of pollution on fish using proteomics can identify proteins that respond to pollutants, aiding in ecological conservation efforts.

Challenges in Proteomics

Proteomics, the large-scale study of proteins, holds immense promise for advancing our understanding of biological systems. However, several key challenges hinder its full potential:

- **Sample Complexity:** The vast differences in protein abundance make it difficult to detect low-level protein and the diverse array of PTMs adds complexity to protein analysis. Also, the multiple forms of a single protein increase the challenge of identification and characterization.
- **Data Acquisition and Analysis:** The sheer volume of data generated necessitates sophisticated computational tools for handling and analysis. Along with this, extracting meaningful insights from complex proteomic datasets remains a significant hurdle. And, ensuring consistent results across different experiments is crucial but challenging.
- **Technical Limitations:** Current technologies may not be sensitive enough to detect low-abundance proteins. Proteomics involves complex techniques and mastering these techniques requires significant expertise and can be technically demanding. Likewise, the lack of standardized protocols hampers data comparability and reproducibility.
- **Biological Complexity:** Understanding protein function requires knowledge of their interactions with other proteins, and analyzing protein localization within cells is crucial but challenging. Protein expression and function vary across time and space, requiring sophisticated analysis.
- **Proteoform Complexity:** The human proteome is incredibly complex, with a vast diversity of proteoforms (different forms of proteins resulting from genetic variations, alternative splicing, and post-translational modifications). This complexity adds layers of difficulty to proteomic analysis and data interpretation

Future Prospectus of Proteomics

Proteomics has the potential to significantly impact various fields, including medicine, agriculture, and biotechnology.

- **Personalized Medicine:** By analysing individual protein profiles, clinical proteomics can enable tailored treatments, improving patient outcomes and minimizing side effects.

- **Unveiling Protein Dynamics:** Single-molecule proteomics provides unprecedented insights into protein behaviour, offering a deeper understanding of biological processes.
- **A Holistic View of Biology:** Integrating proteomics with other omics like genomics and metabolomics provides a more complete picture of biological systems, facilitating the discovery of complex networks.
- **Efficient and Comprehensive Analysis:** High-throughput technologies, such as advanced mass spectrometry, accelerate proteomic research, enabling faster and more thorough investigations.
- **Data-Driven Insights:** Artificial intelligence and machine learning tools analyze vast proteomic datasets, identifying patterns and predicting protein functions.
- **Novel Protein Design:** Advances in protein engineering allow the creation of proteins with tailored functions, with applications in areas like drug development and synthetic biology.

Future research will focus on improving data analysis methods, developing more sensitive detection technologies, and expanding the scope of proteomic studies to encompass a wider range of biological systems.

Case Study In Proteomics

Developing a Vaccine Against *Burkholderia cepacia*: A Reverse Vaccinology Approach

Burkholderia cepacia is a dangerous bacterium that infects people with weakened immune systems, especially those with cystic fibrosis. Its resistance to many antibiotics makes finding a cure difficult.

Objective: This study aimed to create a new type of vaccine against B. cepacia using a computer-based approach called multi-epitope-based reverse vaccinology.

Methods: Researchers sequenced the proteome of *B. cepacia* and identified non-redundant proteins. They then predicted B-cell and T-cell epitopes and constructed a vaccine peptide consisting of 14 epitopes linked together. The vaccine was tested for antigenicity, non-toxicity, and solubility. Molecular docking studies were conducted to assess the binding interactions with immune receptors such as MHC-I, MHC-II, and Toll-like receptor 4 (TLR-4)

Results: The designed vaccine showed promising protective immunity against *B. cepacia* in silico studies. The binding free energies indicated robust interactions, suggesting potential efficacy.

Conclusion

This study demonstrates the promise of using computer modeling to design vaccines against difficult-to-treat bacteria like *Burkholderia cepacia*. Further experiments are needed to confirm the vaccine's effectiveness in real-world situations.

References

1. Abul-Husn, N. S., & Kenny, E. E. (2019). Personalized Medicine and the Power of Electronic Health Records. Cell, 177(1), 58–69. https://doi.org/10.1016/j.cell.2019.02.039

2. Ahmadian, A., Ehn, M., & Hober, S. (2006). Pyrosequencing: History, biochemistry and future. Clinica Chimica Acta, 363(1–2), 83–94. https://doi.org/10.1016/j.cccn.2005.04.038
3. Altschul, S. F., Gish, W., Miller, W., Myers, E. W., & Lipman, D. J. (1990). Basic local alignment search tool. Journal of Molecular Biology, 215(3), 403–410. https://doi.org/10.1016/S0022-2836(05)80360-2
4. Anders, S., & Huber, W. (2010). Differential expression analysis for sequence count data. Genome Biology, 11(10), R106. https://doi.org/10.1186/gb-2010-11-10-r106
5. Ayling, M., Clark, M. D., & Leggett, R. M. (2020). New approaches for metagenome assembly with short reads. Briefings in Bioinformatics, 21(2), 584–594. https://doi.org/10.1093/bib/bbz020
6. Babadi, M., Fu, J. M., Lee, S. K., Smirnov, A. N., Gauthier, L. D., Walker, M., Benjamin, D. I., Zhao, X., Karczewski, K. J., Wong, I., Collins, R. L., Sanchis-Juan, A., Brand, H., Banks, E., & Talkowski, M. E. (2023). GATK-gCNV enables the discovery of rare copy number variants from exome sequencing data. Nature Genetics, 55(9), 1589–1597. https://doi.org/10.1038/s41588-023-01449-0
7. Bagger, F. O., Borgwardt, L., Jespersen, A. S., Hansen, A. R., Bertelsen, B., Kodama, M., & Nielsen, F. C. (2024). Whole genome sequencing in clinical practice. BMC Medical Genomics, 17(1), 39. https://doi.org/10.1186/s12920-024-01795-w
8. Bankevich, A., Nurk, S., Antipov, D., Gurevich, A. A., Dvorkin, M., Kulikov, A. S., Lesin, V. M., Nikolenko, S. I., Pham, S., Prjibelski, A. D., Pyshkin, A. V., Sirotkin, A. V., Vyahhi, N., Tesler, G., Alekseyev, M. A., & Pevzner, P. A. (2012). SPAdes: A New Genome Assembly Algorithm and Its Applications to Single-Cell Sequencing. Journal of Computational Biology, 19(5), 455–477. https://doi.org/10.1089/cmb.2012.0021
9. https://www.uniprot.org/
10. http://geneontology.org/
11. https://www.creative-proteomics.com/resource/peptide-mass-fingerprinting-vs-tandem-mass-spectrometry.htm#:~:text=This%20uniqueness%20leads%20to%20the,proteins%20by%20comparing%20the%20measured
12. https://www.genome.jp/kegg/
13. https://string-db.org/
14. https://thebiogrid.org
15. https://www.ebi.ac.uk/pride/
16. https://blast.ncbi.nlm.nih.gov/Blast.cgi
17. https://www.ebi.ac.uk/jdispatcher/msa/clustalo

8 Computer Aided Drug Designing

Introduction

Drug design and discovery form the cornerstone of modern pharmaceutical development, providing innovative solutions to treat diseases and improve human health. This multidisciplinary process integrates biology, chemistry, and computational science to identify and optimize compounds capable of interacting with biological targets. Over the decades, advances in technology and an increased understanding of molecular biology have transformed drug discovery from serendipitous observations to rational, systematic design approaches.

Drug design and discovery involve identifying and optimizing chemical compounds to serve as effective therapeutic agents. This process has evolved significantly from traditional trial-and-error methods to rational, computer-aided approaches. Computer-aided drug design (CADD) has emerged as an indispensable tool in contemporary pharmaceutical research and development. This chapter provides an in-depth exploration of the principles, methodologies, and applications of CADD in the discovery and optimization of bioactive compounds. It delves into the computational techniques, molecular modeling, and simulation methods used in rational drug design, outlining their significance in accelerating the drug development process. Furthermore, this chapter discusses the challenges, limitations, and future prospects of CADD in the pharmaceutical industry.

Definition and Concept of Drug Design

Drug design is the thoughtful and strategic process of creating molecules that can interact with specific biological targets to treat or manage diseases. Here's how it works:

- **Identifying a Target:** Scientists first pinpoint a biological component, such as an enzyme, receptor, or gene, that plays a key role in a disease.
- **Designing Molecules:** Once the target is identified, researchers design or discover molecules (like small chemicals, peptides, or biological agents) that can interact with it effectively.
- **Optimizing Leads:** These molecules are then fine-tuned to ensure they work well, are safe for the body, and have minimal side effects.

Drug Design Generally Falls into Two Categories

- **Structure-Based Drug Design (SBDD):** This method uses the 3D structure of the target (like a protein) to design drugs that fit it like a key in a lock.
- **Ligand-Based Drug Design (LBDD):** Here, researchers use existing molecules known to be effective to inspire new drug designs.

Overview of the Drug Discovery Pipeline

The journey from idea to medication is a detailed, step-by-step process that evolves over several phases:

A. Target Identification and Validation

- **What's the Goal?** To discover biomolecules, like proteins or genes, that are involved in disease progression.
- **How's it Done?** Techniques like genomics (studying genes), proteomics (studying proteins), and bioinformatics (using computer-based tools) help identify these targets.
- **Validation:** Scientists confirm the target's role through lab experiments and animal studies.

B. Hit Identification and Lead Optimization

- **Finding the Starting Point:** Researchers use methods like high-throughput screening (testing many compounds quickly) or computer simulations to identify potential "hits"—molecules that show initial promise.
- **Making It Better:** These hits are refined to improve their effectiveness and ensure they are safe and stable, turning them into lead compounds ready for further testing.

C. Preclinical Development

Purpose: This stage focuses on testing the safety, toxicity, and effectiveness of the lead compounds in laboratory and animal models.

Outcome: Only compounds with strong safety profiles move forward to human trials.

D. Clinical Trials

Clinical trials involve testing the drug on people and are divided into three key phases:

- **Phase I:** Tests for safety and appropriate dosage using a small group of healthy individuals.
- **Phase II:** Evaluates the drug's effectiveness and side effects in a larger group of patients.
- **Phase III:** Confirms the drug's efficacy in large-scale studies while closely monitoring for adverse reactions.

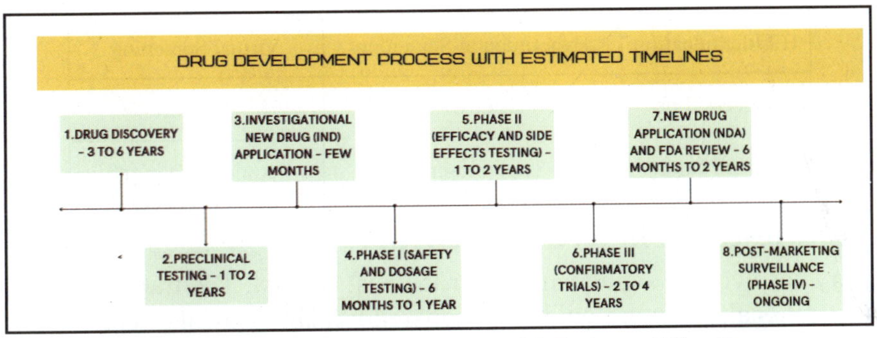

Figure 1: Drug Development process with Estimated Timelines

E. Regulatory Approval

Once clinical trials are successful, the drug's data is submitted to regulatory bodies like the FDA (United States) or EMA (Europe) for review and approval. Even after approval, post-marketing surveillance (Phase IV) ensures the drug's long-term safety and effectiveness are continuously monitored.

By humanizing this process, we see it as a blend of science, creativity, and care, all aiming to improve the quality of life for patients worldwide.

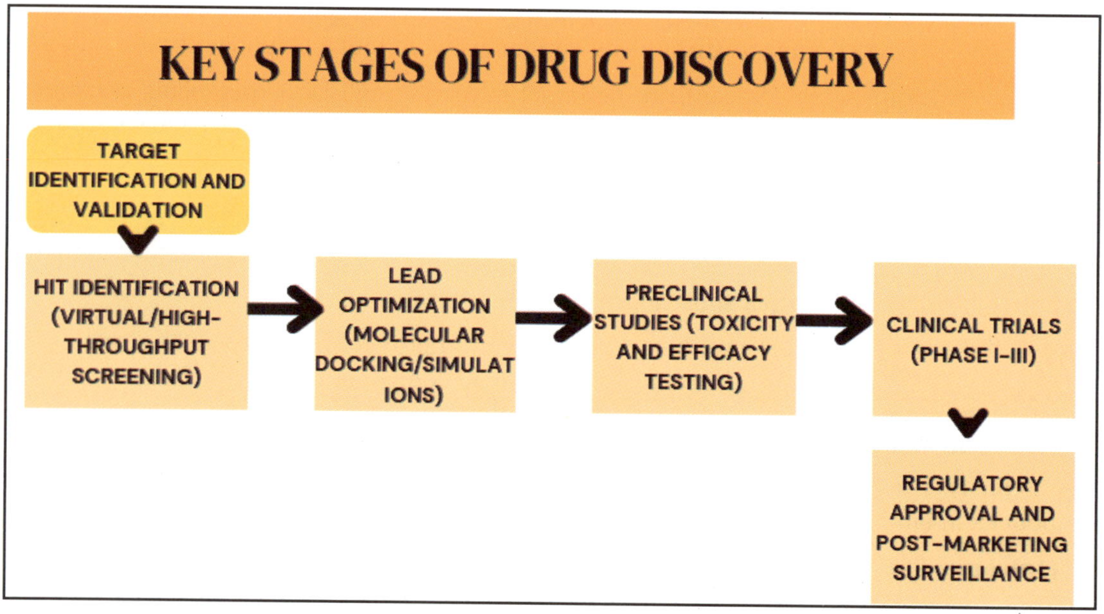

Figure 2: Key Stages of Drug Discovery

Table 1: Comparison of Traditional vs. Computer-Aided Drug Design (CADD)

Feature	Traditional Drug Discovery	Computer-Aided Drug Design
Time Required	10–15 years	5–10 years
Cost	High	Reduced
Hit Identification	Experimental Screening	Virtual Screening
Success Rate	Low	Higher
Optimization	Trial-and-error	Rational Design

Historical Milestones in Drug Discovery

Drug discovery has evolved significantly over the past century. The transition from empirical approaches to rational, structure-guided design reflects the growing sophistication of drug discovery methodologies.

Table 2: Key Milestones in Drug Discovery

Year	Milestone	Impact
1928	Discovery of Penicillin	First antibiotic; transformed infectious disease treatment
1960	First Use of QSAR in Drug Design	Paved the way for computational drug discovery
1971	Discovery of Aspirin's Mechanism (COX Inhibition)	Landmark in understanding drug mechanisms
1980	FDA Approval of Captopril	First ACE inhibitor; rational drug design success
1990	Introduction of Structure-Based Drug Design (SBDD)	Enabled molecular docking and protein-targeted therapies
1997	FDA Approval of Rituximab	First monoclonal antibody therapy for cancer
2001	Imatinib Approval (Gleevec)	First targeted cancer therapy
2013	CRISPR Gene-Editing for Drug Discovery	Revolutionized genetic drug target identification
2020	AI-Driven SARS-CoV-2 Drug Discovery	Accelerated identification of potential inhibitors
2023	First AI-Designed Drug Enters Clinical Trials	Marked new era of AI-driven de novo drug design

Table 3: Overview of Software Tools Used in SBDD and LBDD

Category	Software Tool	Description	Application
Visualization	PyMOL	A molecular visualization tool for analyzing 3D structures of proteins and ligands.	Viewing protein structures, binding site visualization.

Molecular Docking	AutoDock	A docking tool for predicting how small molecules bind to a target protein.	Docking studies, scoring binding affinities.
	CB-Dock	A web-based docking tool with automated binding site prediction.	Simplified docking workflows for non-experts.
Structural Analysis	Chimera	Visualization and analysis of molecular structures and interactions.	Analyzing protein-ligand complexes.
	MOE (Molecular Operating Environment)	Integrated software for molecular modeling and design.	Docking, pharmacophore modeling, QSAR analysis.
Virtual Screening	Schrodinger Glide	High-performance docking software for virtual screening and lead optimization.	Screening virtual libraries for potential drug candidates.
	GOLD	Genetic optimization-based docking software.	Flexible docking of ligands into protein binding sites.
Fragment-Based Design	OpenEye's ROCS	Rapid overlay of chemical structures to identify fragment similarities.	Fragment-based drug design, similarity searches.
Homology Modeling	SWISS-MODEL	A web-based tool for building protein models from homologous sequences.	Predicting protein structures when no experimental data is available.

Pharmacophore Modeling	LigandScout	A tool for creating and visualizing pharmacophore models.	Identifying critical features required for target binding.
QSAR Modeling	ChemAxon's JChem	Software suite for chemical database management and QSAR modeling.	Predicting biological activity based on chemical structure.
	KNIME	Open-source platform for data analytics and QSAR workflows.	Automating QSAR analysis and lead optimization.
Machine Learning & AI	DeepChem	Machine learning platform for drug discovery and material science.	Enhancing virtual screening and lead discovery with AI models.
Molecular Dynamics	GROMACS	Software for simulating the molecular dynamics of proteins and ligands.	Studying stability and interactions in protein-ligand complexes.
	AMBER	Molecular dynamics software for biomolecules.	Simulating dynamic interactions and refining docking results.
Database Integration	Biovia Discovery Studio	Suite for modeling, simulation, and database integration.	Connecting SBDD/LBDD workflows with structural and compound databases.
Ligand Library Preparation	RDKit	Open-source toolkit for cheminformatics and molecule preparation.	Preparing chemical libraries for docking and screening.

	Open Babel	Tool for converting file formats and preparing ligands for simulations.	Ensuring compatibility across software tools.

Structure Based Drug Designing (SBDD)

Structure-Based Drug Design (SBDD) focuses on creating drugs using the 3D structure of a biological target, often a protein. This approach depends on high-resolution structural data obtained through advanced techniques like X-ray crystallography, NMR spectroscopy, or cryo-electron microscopy (cryo-EM).

Key Steps in SBDD

A.Target Selection and Validation:

- Identify a target protein, enzyme, or receptor involved in the disease.
- Confirm the target's role in the disease through experimental validation.

B. Structural Determination:

- Obtain the target protein's 3D structure.
- Use resources like the Protein Data Bank (PDB) to access structural information.

C. Binding Site Identification:

- Use computational tools to analyze potential active or allosteric sites on the target.
- Hotspot mapping highlights key regions where ligands are most likely to bind.

D. Molecular Docking and Screening:

- Simulate how small molecules fit into the target's binding pocket.
- Screen virtual libraries of compounds and rank their binding affinities to identify promising candidates.

E. Lead Optimization:

- Modify promising molecules to improve their binding strength, specificity, and safety profile.
- Refine compounds through cycles of docking and molecular simulations.

Techniques in SBDD

- **Molecular Docking:** Predicts how potential drugs bind to the target protein.
- **Homology Modeling:** Builds 3D models of proteins using similar, known structures.
- **Fragment-Based Drug Design (FBDD):** Combines small molecular fragments into potent drug candidates.

Applications of SBDD

(a) Cancer Therapy

- Development of kinase inhibitors like Imatinib for chronic myeloid leukemia.
- Discovery of EGFR inhibitors (e.g., Erlotinib, Gefitinib) for non-small cell lung cancer.

(b) Antiviral Drug Development

- Creation of protease inhibitors for HIV (e.g., Saquinavir, Ritonavir).
- Design of RNA polymerase inhibitors for Hepatitis C and SARS-CoV-2.

(c) Antibacterial Agents

- Discovery of inhibitors targeting bacterial DNA gyrase and topoisomerase IV (e.g., Fluoroquinolones).
- Design of antibiotics targeting bacterial ribosomes or cell wall synthesis.

(d)Neurodegenerative Diseases

- Development of inhibitors targeting beta-amyloid aggregation in Alzheimer's disease.
- Creation of molecules modulating kinases and other enzymes involved in Parkinson's disease.

(e) Cardiovascular Disorders

- Design of renin inhibitors (e.g., Aliskiren) to manage hypertension.
- Development of drugs targeting platelet aggregation for thrombotic disorders.

(f) Immunological Diseases

- Discovery of JAK inhibitors (e.g., Tofacitinib) for rheumatoid arthritis and psoriasis.
- Design of small molecules targeting immune checkpoint proteins (e.g., PD-1, CTLA-4).

(g) Metabolic Disorders

- Development of DPP-4 inhibitors (e.g., Sitagliptin) for diabetes management.
- Design of molecules targeting PPAR receptors for lipid and glucose metabolism disorders.

(h) Rare Genetic Diseases

- Discovery of small molecules targeting mutated proteins in cystic fibrosis (e.g., Ivacaftor).
- Design of chaperones stabilizing defective enzymes in lysosomal storage disorders.

(i) Antifungal Agents

- Design of inhibitors targeting fungal sterol biosynthesis pathways (e.g., Azoles).
- Discovery of molecules that inhibit fungal cell wall components like beta-glucans.

(j)Vaccine Adjuvant Design

- Development of small molecules that enhance immune responses by targeting specific receptors or pathways.

Ligand Based Drug Design

Ligand-Based Drug Design (LBDD) takes a different route, relying on insights from existing molecules known to interact with the target. Unlike SBDD, LBDD doesn't need detailed structural information about the target. Instead, it uses the chemical properties of active molecules to guide the design of new drugs.

Table 4: List of Drugs Discovered Through LBDD

Drug Name	**Indication**	**Discovery Approach**
Penicillin	Bacterial infections	Serendipitous discovery.
Imatinib	Chronic myeloid leukemia	Rational ligand design targeting BCR-ABL kinase.
Ibuprofen	Pain and inflammation	Ligand-based screening of NSAIDs.

Key Steps in LBDD

(a) Data Collection of Known Ligands

- Gather information about molecules already known to interact with the target.
- Use databases like ChEMBL or PubChem to find ligand data.

(b) Quantitative Structure-Activity Relationships (QSAR)

- Build mathematical models to connect a compound's chemical structure with its biological activity.
- Use these models to predict the activity of new compounds.

(c) Pharmacophore Modeling

- Identify key molecular features (e.g., hydrogen bond donors or hydrophobic areas) required for activity.
- Design new ligands that match these essential features.

(d) Virtual Screening and Database Mining

- Search for molecules in virtual libraries that share similar features with known active compounds.
- Use tools like AI and machine learning to enhance the screening process.

(e) Lead Optimization

- Refine the identified compounds to boost their effectiveness and reduce side effects.

Techniques in LBDD

- QSAR Modeling: Predicts how well a compound will work based on its structure.
- Pharmacophore Mapping: Pinpoints the molecular features essential for activity.
- Similarity Searching: Identifies new drug candidates by finding molecules similar to known ones.

Applications of LBDD

(a) Development of non-steroidal anti-inflammatory drugs (NSAIDs)

- LBDD has been pivotal in designing NSAIDs like ibuprofen and naproxen, targeting cyclooxygenase (COX) enzymes to reduce inflammation and pain.

(b) Design of antihistamines

- LBDD enabled the discovery of cetirizine and loratadine, selectively targeting histamine H1 receptors to treat allergic conditions.

(c) Antiviral Drug Discovery

- LBDD contributed to the development of drugs like zanamivir (Relenza) and oseltamivir (Tamiflu), targeting neuraminidase enzymes to combat influenza.
- Recently, remdesivir and other antivirals have been optimized for viral RNA polymerase inhibition through LBDD approaches.

(d) Cancer Therapy

- Ligand-based techniques facilitated the design of tyrosine kinase inhibitors (TKIs) such as erlotinib and gefitinib, targeting epidermal growth factor receptors (EGFR) in non-small cell lung cancer.

(e) Cardiovascular Drug Development

- Beta-blockers (e.g., propranolol) and angiotensin II receptor blockers (ARBs), such as losartan, were optimized using ligand-based methods to manage hypertension and heart failure.

(f) CNS (Central Nervous System) Drugs

- Development of dopamine receptor agonists (e.g., pramipexole) for Parkinson's disease and serotonin reuptake inhibitors (SSRIs) (e.g., fluoxetine) for depression and anxiety.

(g) Antibacterial Agents

- Discovery of quinolone antibiotics like ciprofloxacin, which inhibit bacterial DNA gyrase and topoisomerase IV, benefitted from LBDD approaches.

(h) Diabetes Treatment

- LBDD was essential in designing dipeptidyl peptidase-4 (DPP-4) inhibitors like sitagliptin to improve glycemic control in type 2 diabetes patients.

(i) Opioid Receptor Modulators

- Ligand-based techniques were instrumental in optimizing compounds like buprenorphine and naloxone, targeting opioid receptors for pain management and opioid addiction.

Table 5: SBDD vs. LBDD

Feature	Structure-Based Drug Design (SBDD)	Ligand-Based Drug Design (LBDD)
Requirement	3D structure of the target	Known active ligands
Primary Approach	Docking and molecular modeling	QSAR, pharmacophore modeling
Structural Data	Essential	Not required
Screening Technique	Virtual docking	Ligand similarity screening
Lead Optimization	Iterative structure refinement	Ligand modification
Applications	Kinase inhibitors, protease inhibitors	NSAIDs, antihistamines

Computer-Aided Drug Design (CADD) represents a revolutionary approach to drug discovery and development, leveraging computational tools to streamline the traditionally laborious and expensive process. This chapter delves into the workflows, methodologies, and tools that define CADD and its subfields, supported by real-world case studies and practical applications.

Table 6: Tools and Their Applications in CADD

Category	Tool	Application
Docking Software	AutoDock	Molecular docking and binding affinity prediction.
	CB-Dock	Simplified docking with automated binding site detection.
Visualization Tools	PyMOL	3D visualization of protein-ligand interactions.
	Chimera	Structural analysis and interaction visualization.

ADME Prediction	SwissADME	Drug-likeness, ADME, and bioavailability prediction.
	ADMETlab	Advanced prediction of ADME and toxicity profiles.

Molecular Modeling Techniques

Molecular Docking

Molecular modeling encompasses a suite of computational methods used to simulate molecular structures and interactions. These techniques are foundational to Computer-Aided Drug Design (CADD), aiding in the discovery and optimization of therapeutic compounds.

Molecular docking is a critical computational technique in drug discovery, used to predict how a small molecule, such as a drug candidate, binds to a biological target like a protein or an enzyme. By simulating the interactions between a ligand (a potential drug) and a receptor (the target molecule), molecular docking provides insights into the binding affinity and orientation of the ligand within the active site of the receptor. This approach not only aids in understanding molecular interactions but also helps prioritize compounds for further experimental validation. Docking algorithms are central to this process and vary in complexity and accuracy. For example, **AutoDock**, one of the most widely used tools, employs a flexible docking methodology with energy-based scoring functions to evaluate ligand binding. It balances computational efficiency with predictive accuracy, making it a preferred choice in virtual screening campaigns. Another prominent tool, **GOLD**, utilizes a genetic algorithm to explore ligand binding, offering enhanced flexibility and precise scoring for diverse chemical libraries.

The interactions between the ligand and receptor are driven by fundamental chemical forces. Hydrogen bonds play a crucial role in anchoring the ligand to the receptor's active site, providing specificity and stability. Hydrophobic interactions facilitate the positioning of non-polar regions of the ligand in compatible regions of the receptor, while van der Waals forces contribute to the overall stability and complementarity of the binding interaction. These interactions collectively determine the binding orientation and strength, which are quantified through scoring functions. Scoring functions are mathematical models used to estimate the binding affinity between a ligand and its receptor. They evaluate parameters such as binding energy, the fit of the ligand within the receptor's binding pocket, and the contribution of various molecular interactions. By providing a numerical score, these functions enable researchers to rank potential drug candidates, prioritizing those with the most promising binding profiles.

Molecular docking has wide-ranging applications in drug discovery. **Virtual screening** is a prominent use case, where large chemical libraries are docked computationally to identify compounds with high binding affinity to a target protein. This high-throughput approach significantly reduces the number of candidates requiring experimental testing, accelerating the drug discovery process. Another key application is **lead optimization**, where molecular docking is used iteratively to refine the chemical structure of a lead compound, enhancing its potency, selectivity, and pharmacokinetic properties. Overall, molecular docking serves as an indispensable tool in modern drug discovery, bridging the gap between computational predictions and experimental validation. Its ability to simulate molecular interactions and rank potential candidates has transformed the traditional approach to drug design, making it more targeted, efficient, and cost-effective.

Table 7: Comparison of Homology Modeling and Docking Tools

Aspect	Homology Modeling Tools	Docking Tools
Purpose	Predict 3D structure of proteins.	Simulate ligand binding.
Key Tools	SWISS-MODEL, MODELLER	AutoDock, GOLD
Input Requirement	Protein sequence	Protein structure, ligand structure.
Output	Protein 3D model	Docking poses and binding scores.

Molecular Dynamics Simulation

Molecular dynamics (MD) simulations are a powerful computational technique used to explore the behavior of molecular systems over time, providing a dynamic perspective on atomic interactions that static structures cannot reveal. By solving Newton's equations of motion for each atom in the system, MD simulations capture the intricate movements and conformational changes of molecules under specified conditions, such as temperature and pressure. This approach enables researchers to analyze the stability, flexibility, and interactions of complex biomolecular systems, including ligand-receptor complexes, in a time-resolved manner.

The fundamental principle of MD lies in its ability to track atomic motions in real-time. These simulations begin with an initial atomic configuration, often derived from experimental structures such as those obtained via X-ray crystallography or NMR spectroscopy. The system is then subjected to defined physical conditions and evolved through iterative calculations of forces and accelerations. This time-resolved tracking provides insights into molecular behavior, including the stability of ligand-receptor interactions, conformational changes in proteins, and the pathways through which ligands bind or dissociate. Central to MD simulations is the concept of force fields, which are mathematical models that describe the potential energy surfaces governing atomic interactions. Force fields include parameters for bond stretching, angle bending, torsional rotations, and non-bonded interactions such as van der Waals forces and electrostatics. Popular force fields like AMBER, CHARMM, and GROMOS are widely used in biomolecular simulations due to their accuracy and robustness. These force fields play a critical role in determining the reliability of simulation results, as they influence the calculated dynamics and energetics of the system. MD simulations have several significant applications in drug discovery and molecular biology. One important application is the study of protein conformational changes. Proteins are inherently dynamic molecules, and their functions often depend on their ability to adopt different conformations. MD simulations allow researchers to investigate these conformational transitions, such as the opening and closing of active sites, which are critical for understanding enzyme mechanisms or receptor activation. This dynamic perspective helps identify potential binding sites and allosteric regions that may not be apparent from static structures.

Another key application of MD simulations is in analyzing ligand binding kinetics. By simulating the interactions between a ligand and its receptor over time, MD provides valuable insights into the stability of the complex, binding pathways, and intermediate states. These simulations can reveal how ligands navigate the receptor's binding pocket, the energetics of binding and unbinding, and the role of water molecules or ions in facilitating interactions. Such insights are invaluable for optimizing lead compounds and designing drugs with enhanced binding affinities and specificity. Molecular dynamics simulations also contribute to understanding the effects of mutations, predicting protein stability, and exploring macromolecular assemblies such as membrane proteins or nucleic acid-protein complexes. By capturing the temporal evolution of molecular systems, MD bridges the gap between static structural biology and experimental biophysics, offering a comprehensive view of molecular interactions that underpin biological function and drug efficacy. This makes it an indispensable tool in modern computational drug design and molecular research.

Computer-Aided Drug Design (CADD) through Machine Learning and AI

The integration of machine learning (ML) and artificial intelligence (AI) into computer-aided drug design (CADD) has revolutionized the pharmaceutical industry. These technologies bring data-driven precision, speed, and innovation to drug discovery, enabling the analysis of vast datasets, prediction of molecular interactions, and optimization of lead compounds with unprecedented accuracy. The convergence of CADD with ML and AI is reshaping traditional drug development paradigms, making the process more cost-effective and efficient. Machine learning involves the use of algorithms that learn from data to make predictions or decisions without explicit programming. AI extends this by simulating human-like intelligence for problem-solving and decision-making. In CADD, ML and AI contribute to tasks such as:

- Predicting drug-target interactions.
- Identifying potential side effects or toxicity.
- Designing novel molecules with desirable properties.
- Enhancing virtual screening processes.

By leveraging advanced algorithms, such as neural networks, support vector machines, and random forests, these technologies enable faster and more accurate identification of drug candidates.

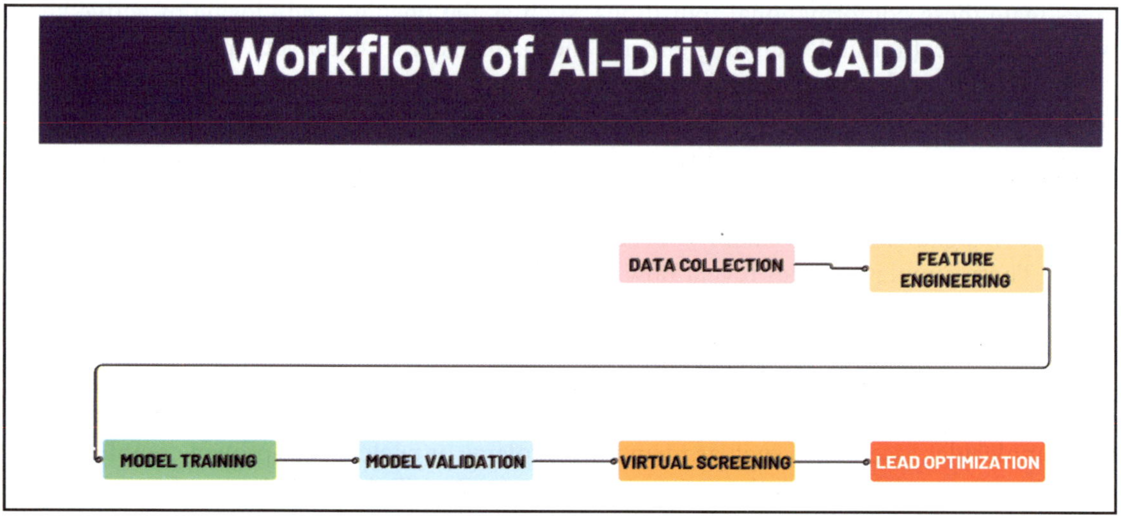

Figure 3: Workflow of AI Driven CADD

Applications of ML and AI in CADD

(a) Virtual Screening

Virtual screening involves evaluating large libraries of compounds to identify those most likely to bind to a biological target. ML models are trained on datasets of known ligands and non-ligands to predict binding affinities, reducing the time and resources required for experimental testing. AI-powered platforms can process millions of compounds efficiently, ranking candidates by their potential efficacy.

(b) De Novo Drug Design

AI facilitates the generation of novel molecular structures with specific properties. Generative adversarial networks (GANs) and recurrent neural networks (RNNs) create chemical structures from scratch, optimizing them for factors like binding affinity, solubility, and pharmacokinetics. This approach accelerates the discovery of unique compounds that might not emerge through traditional methods.

(c) Drug-Target Interaction Prediction

Understanding how a drug interacts with its target is crucial for efficacy and safety. ML models analyze chemical and biological data to predict these interactions, using features such as molecular fingerprints, protein sequences, and structural information. Techniques like deep learning enable the identification of complex patterns in high-dimensional data.

(d) ADME Tox Prediction

Absorption, distribution, metabolism, excretion, and toxicity (ADME-Tox) properties are vital for drug safety and efficacy. ML models trained on experimental ADME-Tox datasets predict these properties for new compounds. Tools like SwissADME and ADMETlab integrate ML algorithms for accurate predictions, reducing the likelihood of late-stage drug failures.

(e) Lead Optimization

AI enhances lead optimization by analyzing molecular structures and suggesting modifications to improve potency, selectivity, and pharmacokinetics. Reinforcement learning algorithms explore chemical space to identify optimal modifications while adhering to specific constraints, such as synthetic feasibility.

(f) Biomarker Discovery and Precision Medicine

AI-driven CADD supports the identification of biomarkers for personalized therapies. By integrating genomic, proteomic, and clinical data, ML models identify patient subgroups most likely to benefit from a given drug, paving the way for precision medicine.

Techniques and Algorithms in ML/AI for CADD

(a) Supervised Learning

Supervised learning algorithms like support vector machines (SVMs) and random forests are widely used in virtual screening and ADME-Tox prediction. These models learn from labeled datasets, making them effective for tasks with well-defined outcomes.

(b) Deep Learning

Deep neural networks (DNNs), convolutional neural networks (CNNs), and graph neural networks (GNNs) have revolutionized drug design by handling complex and unstructured data, such as molecular graphs and protein structures. GNNs, in particular, are effective for predicting molecular properties and interactions.

(c) Natural Language Processing (NLP)

NLP techniques analyze scientific literature, patents, and databases to extract valuable insights about targets, pathways, and potential drug candidates. AI systems like IBM Watson for Drug Discovery utilize NLP for data mining and hypothesis generation.

(d) Reinforcement Learning

In reinforcement learning, models learn through trial and error to optimize molecular structures. This approach is particularly useful in de novo drug design, where AI explores chemical space iteratively to achieve desired properties.

(e) Generative Models

GANs and variational autoencoders (VAEs) create novel compounds by learning the underlying distribution of molecular data. These models generate chemically valid and diverse molecules, enabling rapid exploration of the chemical space.

The Role of Omics in Computer-Aided Drug Design (CADD)

The advent of high-throughput omics technologies has revolutionized the field of drug discovery and development. Omics disciplines, encompassing genomics, transcriptomics, proteomics, metabolomics, and epigenomics, provide comprehensive datasets that elucidate the molecular basis of diseases. By integrating omics data with computer-aided drug design (CADD), researchers can enhance target identification, drug efficacy, and precision medicine strategies.

This chapter explores how omics drives innovation in CADD, enabling more efficient and accurate drug discovery pipelines.

Omics technologies generate large-scale datasets that detail the molecular landscape of biological systems. These datasets provide a foundation for identifying druggable targets, elucidating disease pathways, and predicting therapeutic responses.

(a) Genomics

Genomics focuses on the structure, function, and interaction of genes. Advances in sequencing technologies, such as next-generation sequencing (NGS), have enabled the identification of genetic variations associated with diseases. Key applications of genomics in CADD include:

- Target Identification: Genome-wide association studies (GWAS) highlight genes linked to diseases, enabling the selection of relevant drug targets.
- Precision Medicine: Genomic profiles guide the development of personalized therapies, such as PARP inhibitors for BRCA-mutated cancers.

(b) Transcriptomics

Transcriptomics examines gene expression patterns to understand cellular responses under different conditions. Techniques like RNA sequencing (RNA-Seq) provide insights into upregulated or downregulated genes in disease states. Applications in CADD include:

- Pathway Analysis: Identifying dysregulated pathways that serve as therapeutic targets.
- Drug Repurposing: Matching drug-induced transcriptomic changes with disease profiles to identify potential new indications for existing drugs.

(c) Proteomics

Proteomics analyzes the structure, function, and interactions of proteins, which are the primary targets of most drugs. Mass spectrometry (MS) and protein microarrays enable the identification of differentially expressed proteins in diseases. Applications include:

- Target Validation: Confirming the role of specific proteins in disease mechanisms.
- Interaction Mapping: Understanding protein-protein and protein-ligand interactions to design effective inhibitors.

Example figure:

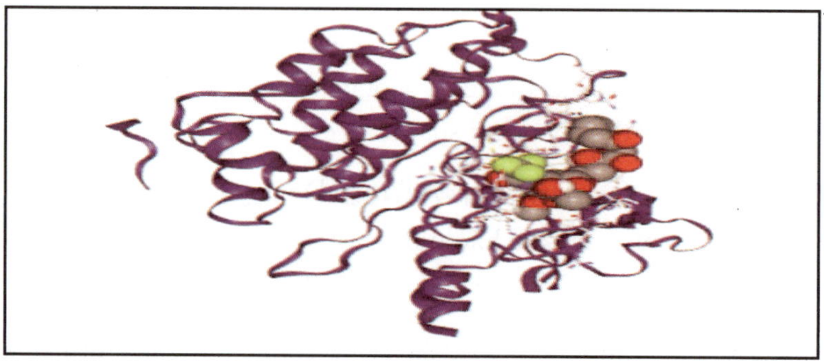

Figure1: Molecular Docking (6UBW), Valrubicin (drug) with score -9

(d) Metabolomics

Metabolomics studies the dynamic profile of metabolites in biological systems. Since metabolites are closely linked to physiological states, their analysis provides direct insights into disease phenotypes. Applications in CADD include:

- Biomarker Discovery: Identifying metabolic biomarkers for disease diagnosis and drug response monitoring.
- **Pathway Reconstruction:** Elucidating metabolic pathways disrupted in diseases for therapeutic intervention.

(e) Epigenomics

Epigenomics explores modifications to the genome that affect gene expression without altering the DNA sequence. Epigenetic changes, such as DNA methylation and histone modification, are implicated in various diseases. Applications in CADD include:

- **Epigenetic Drug Targets:** Development of epigenetic modulators, such as histone deacetylase (HDAC) inhibitors.
- **Predictive Modeling:** Understanding how epigenetic patterns influence drug response.

References

1. Computer Aided Drug Design (CADD): From Ligand-Based Methods to Structure-Based Approaches. (2022). Netherlands: Elsevier Science.
2. Thakur, A., Mehta, V., Nagu, P., Goutam, K. (2024). Computer-Aided Drug Design: QSAR, Molecular Docking, Virtual Screening, Homology and Pharmacophore Modeling. Indonesia: De Gruyter.
3. Molecular Docking for Computer-Aided Drug Design: Fundamentals, Techniques, Resources and Applications. (2021). Netherlands: Academic Press.
4. Computer-Aided Drug Discovery. (2016). Germany: Springer New York.
5. Computer-Aided Drug Design. (2020). Germany: Springer Nature Singapore.
6. Kumar, T. D. A. (2022). Drug Design: A Conceptual Overview. (n.p.): BSP.
7. Vemula D, Jayasurya P, Sushmitha V, Kumar YN, Bhandari V. CADD, AI and ML in drug discovery: A comprehensive review. *Eur J Pharm Sci*. 2023;181:106324. doi:10.1016/j.ejps.2022.106324
8. Chen K, Xu R, Hu X, Li D, Hou T, Kang Y. Recent advances in the development of DprE1 inhibitors using AI/CADD approaches. *Drug Discov Today*. 2024;29(6):103987. doi:10.1016/j.drudis.2024.103987
9. Niazi SK, Mariam Z. Computer-Aided Drug Design and Drug Discovery: A Prospective Analysis. *Pharmaceuticals (Basel)*. 2023;17(1):22. Published 2023 Dec 22. doi:10.3390/ph17010022
10. Fatima I, Rehman A, Ding Y, et al. Breakthroughs in AI and multi-omics for cancer drug discovery: A review [published correction appears in Eur J Med Chem. 2024 Dec 5;279:116968. doi: 10.1016/j.ejmech.2024.116968]. *Eur J Med Chem*. 2024;280:116925. doi:10.1016/j.ejmech.2024.116925
11. Razia IT, Kanwal A, Riaz HF, et al. Recent Trends in Computer-aided Drug Design for Anti-cancer Drug Discovery. *Curr Top Med Chem*. 2023;23(30):2844-2862. doi:10.2174/0115680266258467231107102643

9 Artificial Intelligence and Machine Learning

Introduction

Artificial Intelligence (AI) and Machine Learning (ML) are revolutionizing the biological sciences, offering powerful tools to analyze vast and complex datasets. From decoding genomes to predicting protein structures, these technologies have reshaped how biological research is conducted. AI encompasses systems that mimic human cognitive functions, while ML enables machines to learn from data and improve predictions over time. These advancements align with the explosive growth of data in biology, including genomic sequences, proteomics datasets, and biomedical imaging. Understanding AI and ML through a biological lens involves leveraging computational power to uncover patterns and relationships in data, facilitating breakthroughs in diagnostics, drug discovery, and personalized medicine. Artificial Intelligence (AI) and Machine Learning (ML) are the cornerstone technologies of modern computational biology, driving innovations in genomics, proteomics, and systems biology.

AI is a branch of computer science focused on creating systems capable of performing tasks that typically require human intelligence, such as reasoning, learning, problem-solving, perception, and language understanding. In technical terms, AI encompasses algorithms and models designed to simulate cognitive processes. ML is a subset of AI that uses statistical techniques to give computers the ability to learn from and make decisions or predictions based on data without being explicitly programmed. It involves methods such as supervised learning (e.g., classification, regression), unsupervised learning (e.g., clustering, dimensionality reduction), and reinforcement learning. There is one more term, known as deep learning (DL). DL is a subset of ML that utilizes artificial neural networks with multiple layers to model complex patterns in data. It excels in tasks such as image recognition, speech processing, and natural language understanding due to its ability to automatically extract features from raw data.

In biological sciences, AI/ML technologies address challenges related to the sheer scale and complexity of data. For example- **AI** is applied in designing algorithms to predict gene-disease associations or simulate biochemical pathways, **ML** powers tasks such as clustering single-cell RNA-Seq data or identifying protein families, **DL** models, such as convolutional neural networks, are transforming biomedical image analysis, from histopathology to microscopy. To understand the hierarchy, AI can be viewed as the broad field, encompassing ML as a subset. In turn, ML encompasses DL as a specialized subset.

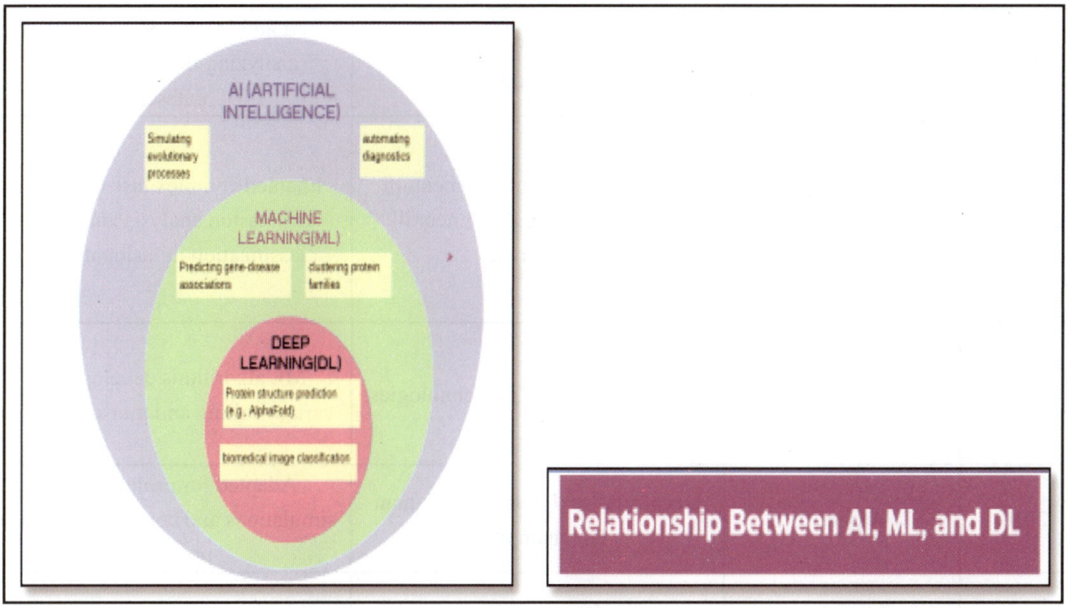

Figure 1: Illustrating the relationship between AI, ML, and DL, along with its biological relevance

Table1: Historical Milestones in AI and Mach ine Learning in Biology

Year	Event	Description	Impact on Biology
1950	Turing Test	Proposed by Alan Turing to evaluate machine intelligence.	Sparked the idea of using computational systems to simulate biological intelligence.
1956	Dartmouth Conference	The term "Artificial Intelligence" was coined.	Laid the foundation for AI applications, including bioinformatics.
1970s	Expert Systems	Development of systems like Dendral for chemical analysis.	Used for interpreting mass spectrometry data in biochemistry.
1980s	Neural Networks	Introduction of backpropagation algorithms for training neural networks.	Early applications in protein folding and structure prediction.
1990	BLAST Tool	Basic Local Alignment Search Tool introduced.	Revolutionized sequence alignment and comparison in genomics.
1996	IBM's Deep Blue	First AI system to beat a human chess champion.	Inspired algorithms for computational biology problems like drug design.

2001	Human Genome Project	Completion of the first human genome sequence.	AI and ML became critical in analyzing massive genomic datasets.
2006	Deep Learning	Deep learning gains momentum with breakthroughs in neural networks.	Enabled complex tasks like gene expression analysis and image classification in histopathology.
2010	High-Throughput Sequencing	Advancement in NGS technologies.	ML algorithms developed for variant calling and transcriptomics.
2016	AlphaGo	DeepMind's AI defeated a human Go champion.	Adapted for protein folding simulations and complex biological problems.
2018	AlphaFold	DeepMind's AI predicts protein structures with high accuracy.	Transformed structural biology, solving decades-old problems.
2020	COVID-19 Research	AI and ML used to study SARS-CoV-2.	Enabled rapid vaccine development and drug repurposing efforts.
2022	Multi-Omics Integration	AI integrates genomic, transcriptomic, and proteomic data.	Advanced precision medicine and systems biology.

Types of Machine Learning

Supervised Learning

Definition: Supervised learning involves training a model on a labeled dataset, where each input is paired with the correct output. The model learns to map inputs to outputs and generalize this mapping to new data.

Applications in Biology

(a). Gene Expression Analysis: Predict whether a given gene is associated with a specific disease by training on labeled datasets containing gene-disease relationships.

(b). Protein Classification: Classify proteins into functional groups based on their sequence or structural features.

(c). Disease Diagnosis: Use labeled histopathology images to train models that identify cancerous vs. non-cancerous cells.

Unsupervised Learning

Definition: Unsupervised learning works with unlabeled data, discovering hidden patterns or structures within the dataset without predefined categories.

Applications in Biology

(a). **Single-Cell RNA-Seq Analysis**: Clustering cells based on their gene expression profiles to identify cell types or states.

(b). **Sequence Clustering**: Group similar DNA, RNA, or protein sequences to identify families or evolutionary relationships.

(c). **Metagenomics**: Analyze microbial communities by clustering DNA fragments from environmental samples to classify species.

Semi-Supervised Learning

Definition: Semi-supervised learning combines a small amount of labeled data with a large amount of unlabeled data to improve learning accuracy.

Applications in Biology

(a). **Functional Genomics**: Use labeled genes with known functions to predict the functions of unlabeled genes.

(b). **Drug Response Prediction**: Combine sparse clinical trial data with abundant unlabeled patient records to predict drug efficacy.

(c). **Disease Subtyping**: Identify subtypes of diseases using a small set of annotated patient samples and a larger set of unannotated samples.

Reinforcement Learning

Definition: Reinforcement learning involves training agents to make a sequence of decisions by interacting with an environment. The agent learns from rewards and penalties to maximize its cumulative reward.

Applications in Biology

(a). **Protein Folding Simulations**: Use reinforcement learning to explore protein conformations and predict the most stable structure.

(b). **Synthetic Biology**: Design optimized genetic circuits by simulating interactions and feedback loops.

(c). **Drug Discovery**: Optimize drug-binding poses in-silico by training models to find conformations with the best binding affinities.

The Machine Learning Process: A Biological Perspective

Machine learning (ML) involves a systematic process to build models that can extract meaningful insights or make predictions from data. Below is a detailed explanation of each step, illustrated with a **biological example: predicting the presence of a disease using gene expression data**.

Step 1: Data Collection

The first step is to gather relevant data for the problem at hand.

- **Example**: Collect gene expression data from patients, including both healthy individuals and those diagnosed with a particular disease (e.g., cancer). Sources could include public repositories like GEO (Gene Expression Omnibus) or TCGA (The Cancer Genome Atlas).

Challenges

- Ensuring sufficient data volume for reliable model training.
- Addressing data diversity to represent different populations.

Step 2: Data Preprocessing

Raw biological data often requires extensive cleaning and transformation before use.

Tasks Involved

- Handle missing values.
- Normalize gene expression levels to account for technical biases.
- Perform feature selection to retain only the most informative genes.
- Encode categorical variables.
- **Example**: Normalize RNA-Seq data to convert raw read counts into comparable expression levels. Filter out low-expression genes and impute missing values to create a clean dataset.

Step 3: Choosing the Right Model

The model choice depends on the type of data and the task (e.g., classification, regression, clustering).

- **Example**: For predicting disease presence (a classification problem), suitable models include logistic regression, support vector machines (SVMs), or random forests. Neural networks can be used for more complex datasets.

Considerations

- Simpler models like SVMs work well for small datasets.
- Complex models like deep learning require large datasets.

Step 4: Training the Model

Training involves feeding the preprocessed data into the chosen model to learn patterns from the input-output pairs.

- **Example**: Train a random forest classifier on gene expression data, where the input is gene expression levels, and the output is the disease label (healthy/diseased).
- **Outcome**: The model learns to differentiate between the gene expression profiles of healthy and diseased samples.

Step 5: Evaluating the Model

Evaluate the model's performance using metrics like accuracy, precision, recall, F1-score, or AUC-ROC.

- **Example**: Split the dataset into training and test sets. Use the test set to calculate performance metrics. For example:
- **Accuracy**: Percentage of correct predictions.

- **AUC-ROC**: Measures the model's ability to distinguish between classes.
- **Goal**: Ensure the model generalizes well to unseen data and doesn't overfit.

Metric	Definition	Biological Example
Accuracy	Proportion of correctly classified instances.	Predict the correct cancer subtype for a patient.
Precision	Proportion of true positives out of predicted positives.	Focuses on reducing false positives in disease diagnosis.
Recall	Proportion of true positives out of actual positives.	Identifying all patients with a disease (high sensitivity).
F1-Score	Harmonic mean of precision and recall.	Balances the need for precision and recall in predicting rare diseases.
AUC-ROC	Measures the model's ability to distinguish between classes.	Used in binary classification tasks like cancer vs. healthy predictions.

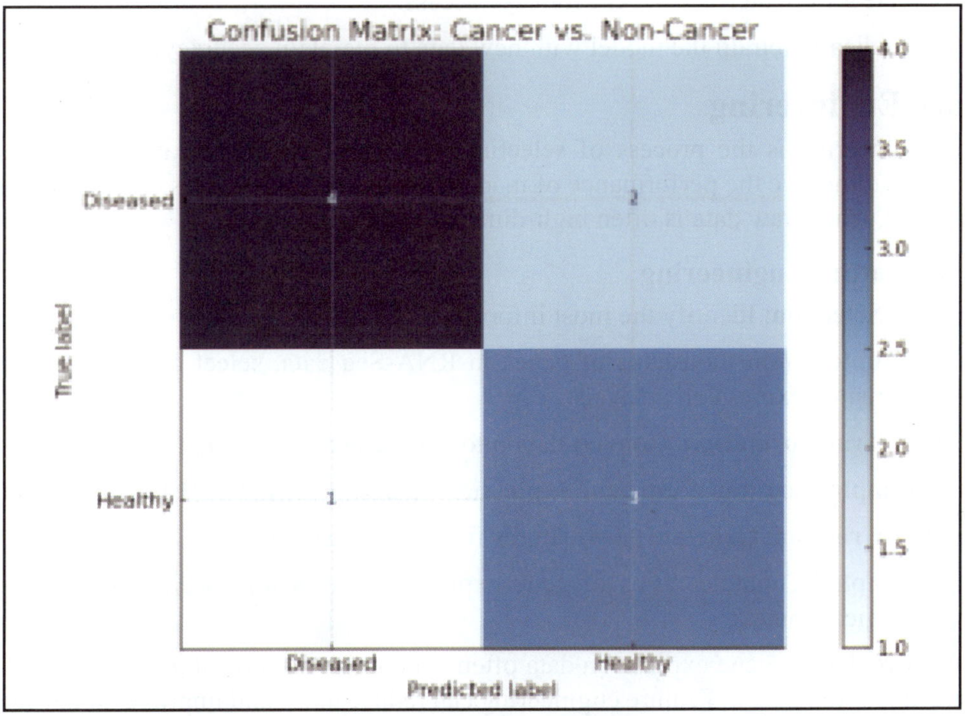

Figure 2: Confusion metrics

- **True Positives (TP):** Correctly identified diseased samples.
- **False Positives (FP):** Healthy samples misclassified as diseased.
- **True Negatives (TN):** Correctly identified healthy samples.
- **False Negatives (FN):** Diseased samples misclassified as healthy.

In a binary classification of cancer vs. non-cancer, the confusion matrix provides insight into model accuracy and potential diagnostic errors.

Step 6: Hyperparameter Tuning and Optimization

Fine-tune the model's hyperparameters to optimize its performance.

- **Example**:Use grid search or random search to optimize parameters like the number of trees in a random forest or the learning rate in a neural network.
- **Outcome**: A model with the best possible performance for the given dataset.

Step 7: Predictions and Deployment

Once the model is trained and optimized, it can be used for making predictions and integrated into real-world systems.

- **Example**:Deploy the trained model in a healthcare setting where clinicians input gene expression data, and the model predicts the likelihood of disease presence. This can assist in early diagnosis and treatment planning.

Deployment Methods

- Create a web interface or API for easy access.
- Periodically retrain the model with new data to maintain accuracy.

Feature Engineering

Feature engineering is the process of selecting, transforming, and creating input variables (features) that improve the performance of machine learning models. It is critical in biological data analysis, where raw data is often high-dimensional and complex.

Steps in Feature Engineering

1. **Feature Selection**: Identify the most informative features.
    - **Example:** From thousands of genes in RNA-Seq data, select the top 500 based on differential expression analysis.
2. **Feature Transformation**: Convert data into a more useful format.
    - **Example:** Log-transform gene expression values to normalize data distribution.
3. **Feature Creation**: Generate new features from existing ones.
    - **Example: C**ompute ratios of gene expression levels for specific pathways to add biological context.

In **cancer prediction**, gene expression data often contains thousands of genes, many of which are irrelevant or redundant. Feature engineering can reduce noise and improve model accuracy:

LASSO (Least Absolute Shrinkage and Selection Operator)

LASSO is a regression analysis method that performs both variable selection and regularization to improve the prediction accuracy of a model. It minimizes the residual sum of squares subject to the sum of the absolute value of the coefficients being less than a fixed value.

Key Features of LASSO

- ❖ Shrinks coefficients of less important features to zero, effectively selecting a subset of features.
- ❖ Helps in handling high-dimensional data.

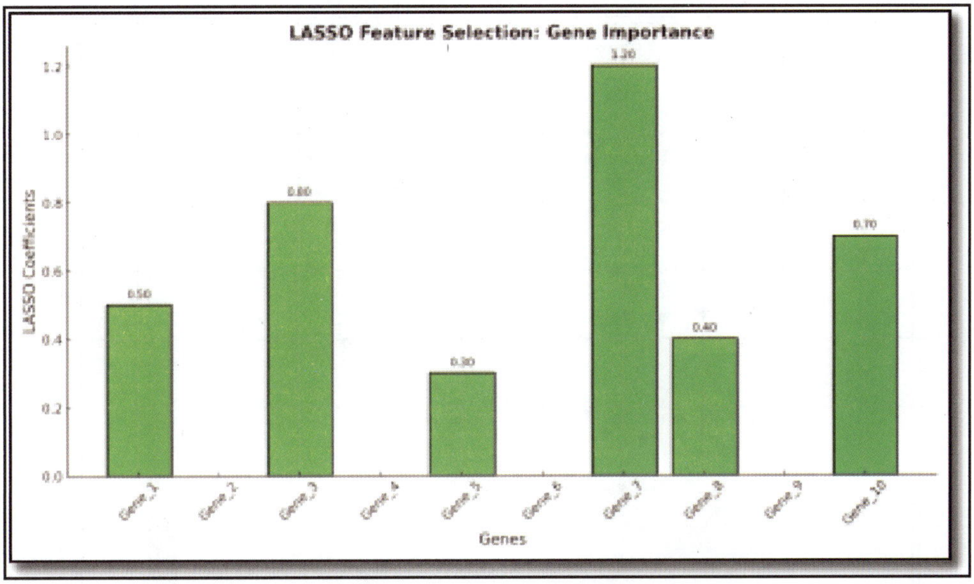

Figure 3: Histogram showing the contribution of each gene

LASSO can identify genes most strongly associated with cancer by penalizing irrelevant or redundant features.

```
# Implementation using python
from sklearn.linear_model import Lasso

# Fit the LASSO model
lasso = Lasso(alpha=0.1)
lasso.fit(X, y)
lasso_coef = lasso.coef_

# Display LASSO coefficients
print("LASSO Coefficients:", lasso_coef)
```

PCA (Principal Component Analysis)

PCA is a dimensionality reduction technique that transforms data into a set of orthogonal components, which capture the maximum variance in the data.

Key Features of PCA

- Reduces the dimensionality of high-dimensional data.
- Facilitates visualization of complex datasets.
- Useful for noise reduction and feature extraction.

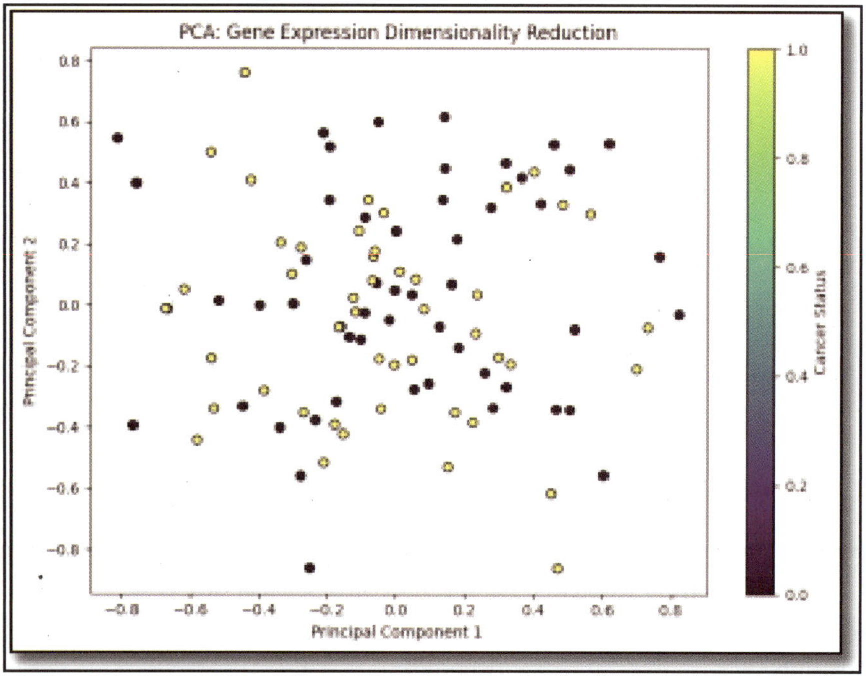

Figure 4: Showing high dimension gene data using PCA

PCA can reduce thousands of gene expression values to a few principal components that capture the most critical patterns, making visualization and analysis easier.

```
from sklearn.decomposition import PCA



# Perform PCA

pca = PCA(n_components=2)

X_pca = pca.fit_transform(X)



# Display explained variance

print("Explained Variance Ratio:", pca.explained_variance_ratio_)
```

Machine Learning and AI Tools in Biology: Python, R, and Specialized Platforms

In the rapidly advancing field of bioinformatics, machine learning (ML) and artificial intelligence (AI) tools are revolutionizing how researchers analyze biological data. From genomic sequencing to protein structure prediction, various programming languages and platforms provide robust solutions to tackle complex biological problems.

A.Python for Predictive Modeling and Sequence Analysis

Python is one of the most widely used languages in computational biology due to its flexibility, extensive libraries, and ease of integration with AI frameworks.

- **TensorFlow & scikit-learn:** TensorFlow facilitates deep learning models for genomic and proteomic data, while scikit-learn offers algorithms for classification, clustering, and regression in predictive modeling.
- **Biopython:** A specialized library for biological computation, Biopython supports tasks like DNA sequence parsing, sequence alignment, and phylogenetics.

Gene Classification Using scikit-learn

```
# biopython must be installed in your system.

from sklearn.model_selection import train_test_split

from sklearn.ensemble import RandomForestClassifier

from sklearn.metrics import accuracy_score

from Bio import SeqIO


# Load DNA sequences (example in FASTA format)

sequences = list(SeqIO.parse("example_sequences.fasta",
"fasta"))

labels = [1 if "cancer" in seq.id else 0 for seq in sequences]


# Feature Extraction (Length of sequence)

features = [len(seq.seq) for seq in sequences]


# Model Training

X_train, X_test, y_train, y_test = train_test_split(features,
labels, test_size=0.3, random_state=42)

classifier = RandomForestClassifier(n_estimators=100)

classifier.fit(np.array(X_train).reshape(-1, 1), y_train)
```

B.R for High-Throughput Sequencing and Model Building

R is another powerful tool, especially for statistical analysis and visualization in bioinformatics.

- **Bioconductor:** This R package is designed for the analysis and comprehension of high-throughput genomic data.
- **caret:** The caret (Classification and Regression Training) package simplifies the process of model training, cross-validation, and feature selection.

```
library(DESeq2)


# Load count data

countData <- read.csv("gene_counts.csv", row.names=1)

colData <- read.csv('sample_info.csv', row.names=1)


# Create DESeq2 dataset

dds <- DESeqDataSetFromMatrix(countData = countData, colData = colData, design = ~ condition)


# Differential expression analysis

dds <- DESeq(dds)

res <- results(dds)


# Filter significant genes

resOrdered <- res[order(res$padj),]

head(resOrdered)
```

C. Specialized Platforms for Computational Biology

Beyond traditional programming, specialized AI platforms like AlphaFold and Rosetta provide highly accurate tools for structural biology.

- **AlphaFold:** DeepMind's AlphaFold predicts 3D protein structures from amino acid sequences with near-experimental accuracy.

Protein Folding with AlphaFold

```
# Install AlphaFold and set up the database

git clone https://github.com/deepmind/alphafold

cd alphafold

bash setup.sh


# Run AlphaFold on a sample FASTA file

python run_alphafold.py --fasta_paths=example.fasta --output_dir=output/
```

Algorithms and Techniques

AI and machine learning (ML) algorithms have opened new frontiers in biology, from disease prediction to drug discovery. Understanding core techniques is essential for applying them to biological datasets.

A. Regression and Classification

- **Logistic Regression**: A foundational algorithm for binary classification. In biology, logistic regression can predict whether a patient has a specific disease based on biomarkers.
- **Support Vector Machines (SVMs)**: SVMs classify data by finding the optimal hyperplane. In genomics, SVMs help classify cancer subtypes based on gene expression profiles.

Example: Predicting the likelihood of diabetes based on patient data (glucose levels, age, BMI).

B. Clustering

- **K-means**: This unsupervised learning algorithm groups biological data into clusters. For example, clustering DNA sequences can identify genetic similarities.
- **Hierarchical Clustering**: Builds a tree of clusters, often used in phylogenetics to map evolutionary relationships between species.

Example: Cluster cancer patients based on gene expression profiles to identify subpopulations with similar responses to treatment.

A. Deep Learning

- **Neural Networks**: Mimics the human brain to detect complex patterns in biological images (e.g., tumor identification in MRI scans).
- **Convolutional Neural Networks (CNNs)**: CNNs excel at image analysis, commonly used for analyzing histopathological slides or detecting cell anomalies.
- **Recurrent Neural Networks (RNNs)**: Ideal for sequential data, RNNs help in analyzing time-series biological data such as EEG signals or protein folding sequences.

B. Ensemble Methods

- **Random Forest**: An ensemble of decision trees, random forests are powerful for classification and feature selection in gene expression data.
- **Gradient Boosting**: Builds models sequentially to correct errors. Applied in drug discovery for predicting molecular interactions.

C. Dimensionality Reduction

- **Principal Component Analysis (PCA)**: Reduces high-dimensional data (e.g., gene expression datasets) to visualize key patterns.
- **t-SNE (t-distributed Stochastic Neighbor Embedding)**: Projects complex biological data onto two dimensions, aiding visualization of cell populations in single-cell RNA sequencing (scRNA-seq).

Ethical Considerations

As AI and ML are increasingly used in biology and healthcare, ethical considerations must be prioritized to ensure fairness, privacy, and responsible AI development.

A. Bias and Fairness AI models can inherit biases from datasets, leading to misdiagnosis or disparities in healthcare. Addressing data imbalance (e.g., underrepresented populations) ensures equitable results.

Example: An AI model trained on European genetic data may underperform for African populations, leading to inaccurate predictions.

B. Privacy Protecting sensitive biological data is crucial. AI models trained on genomic data must comply with privacy regulations like GDPR and HIPAA.

C. Transparency Explainable AI (XAI) techniques allow biologists to interpret model decisions, building trust in AI-assisted diagnostics. For example, highlighting which biomarkers contribute most to a disease prediction.

B. Social Impacts

- **Employment**: AI automates tasks, shifting roles in healthcare and biology.
- **Surveillance**: AI applied to genetic data raises concerns about misuse and genetic profiling.
- **Responsible AI**: Balancing AI advancement with ethical boundaries ensures that its benefits reach all segments of society.

Future Trends

AI and ML are continuously evolving, shaping the future of biology and bioinformatics.

A. Integration with Quantum Computing Quantum computing promises to accelerate complex simulations, such as protein folding and drug interaction modeling, far beyond the capacity of classical computers.

B. Generative AI Generative models like DALL·E and GPT can simulate biological structures, generate synthetic genomic sequences, or create novel drug molecules. AlphaFold exemplifies generative AI's impact on protein structure prediction.

C. Interdisciplinary Research AI's integration across disciplines, from biology to physics and humanities, drives innovation. In synthetic biology, AI designs gene circuits, while in neuroscience, it deciphers brain activity patterns.

D. Evolving Ethical Frameworks As AI becomes more prominent, global collaboration is essential to develop ethical frameworks that guide responsible AI deployment in biology. Initiatives like AI for Good foster transparent and ethical advancements.

A. Genomics and Transcriptomics

Genomics involves the study of an organism's entire DNA, while transcriptomics focuses on RNA expression. AI models can process massive sequencing data, annotate genes, and identify variants critical for disease research.

a. Gene Annotation

Gene annotation identifies coding regions, splice sites, and regulatory elements in raw DNA sequences. Traditional alignment-based methods like BLAST are effective but slow. ML models, however, learn patterns from labeled datasets and generalize to annotate novel genomes.

AI Techniques

- **Hidden Markov Models (HMM):** Models probabilistic transitions between exons and introns.
- **Convolutional Neural Networks (CNN):** Detects motifs and predicts regulatory regions.
- **Recurrent Neural Networks (RNN):** Captures long-term dependencies in DNA sequences.

Hands-on : *Gene Prediction in Bacterial Genomes*

- **Objective:** Annotate genes in a newly sequenced bacterial genome using Python.
- **Dataset:** Download bacterial genomes from NCBI GenBank.
- **Approach**
  1. Preprocess genome data (fasta format).
  2. Use the **GeneMarkS** tool or develop an ML model to predict gene boundaries.
  3. Visualize predicted genes using Biopython.

```
from Bio import SeqIO
import matplotlib.pyplot as plt
# Read genomic data
record = SeqIO.read("bacterial_genome.fasta", "fasta")
plt.plot(range(len(record.seq)), record.seq.count('ATG'))
plt.title("Gene Start Codon (ATG) Frequency Across Genome")
plt.show()
```

b.Variant Calling

Variant calling detects mutations by comparing DNA sequences to a reference genome. AI enhances accuracy by reducing false positives, especially in noisy datasets.

AI Techniques:

- **GATK (Genome Analysis Toolkit):** Uses Bayesian models for SNP and INDEL detection.
- **DeepVariant (Google):** CNN-based variant caller trained on human sequencing data.

c.Single-Cell RNA-Seq (scRNA-seq) Analysis

scRNA-seq profiles gene expression at a single-cell level, uncovering cell heterogeneity in tissues. AI clusters cells into subpopulations, identifying rare cell types.

AI Techniques

- **t-SNE / UMAP:** Dimensionality reduction for visualizing clusters.
- **K-means:** Clustering cell types based on expression profiles.
- Graph Neural Networks (GNN): Identifies cell states and transitions.

Hands-on: *Identifying Immune Cell Subtypes from Tumor Samples*

- **Objective:** Cluster immune cells in tumor microenvironments.
- **Dataset:** Download scRNA-seq datasets from GEO (Gene Expression Omnibus).
- **Approach:**
    1. Normalize data using Scanpy (Python).
    2. Perform clustering using K-means and visualize with t-SNE.
    3. Identify differentially expressed genes in clusters.

```
# Install required Python packages
pip install scanpy scikit-learn matplotlib seaborn pandas
import scanpy as sc
import matplotlib.pyplot as plt
# Step 1: Load single-cell RNA-seq data (example: 10X Genomics)
adata = sc.datasets.pbmc3k()  # Public PBMC dataset (~3k cells)

# Step 2: Basic preprocessing
sc.pp.filter_cells(adata, min_genes=200)  # Remove low-quality cells
sc.pp.filter_genes(adata, min_cells=3)    # Filter out rare genes

# Step 3: Normalize data
sc.pp.normalize_total(adata, target_sum=1e4)  # Scale each cell to 10,000 counts
sc.pp.log1p(adata)  # Log-transform data to reduce skewness

# Optional: Identify highly variable genes
sc.pp.highly_variable_genes(adata, n_top_genes=2000)
adata = adata[:, adata.var['highly_variable']]  # Retain only variable genes

# View normalization results
sc.pl.highly_variable_genes(adata)
# perform clustering
from sklearn.cluster import KMeans
```

```
import numpy as np
# Step 1: Compute PCA for dimensionality reduction
sc.tl.pca(adata, svd_solver='arpack')
X_pca = adata.obsm['X_pca']

# Step 2: Apply K-means clustering
kmeans = KMeans(n_clusters=5, random_state=42)  # 5 clusters
adata.obs['kmeans'] = kmeans.fit_predict(X_pca[:, :10])  # Use first 10 PCs

# Visualize clusters
sc.pl.pca(adata, color='kmeans')

# Visualize with t-SNE
from sklearn.manifold import TSNE

# Compute t-SNE for visualization
adata.obsm['X_tsne'] = TSNE(n_components=2, random_state=42).fit_transform(X_pca[:,
:10])

# Plot t-SNE
sc.pl.embedding(adata, basis='tsne', color='kmeans')
# Identify Differentially Expressed Genes (DEGs)
# Convert kmeans cluster labels to categorical
adata.obs['kmeans'] = adata.obs['kmeans'].astype('category')

# Perform differential gene expression analysis
sc.tl.rank_genes_groups(adata, 'kmeans', method='wilcoxon')
# View top differentially expressed genes
sc.pl.rank_genes_groups(adata, n_genes=20, sharey=False)

# Use genes from your dataset
target_genes = ['TNFRSF4', 'TNFRSF9', 'SUMO3', 'S100B', 'MT-CO1', 'MT-CYB']

# Plot dotplot and violin plot for these genes
```

```
sc.pl.dotplot(adata, var_names=target_genes, groupby='kmeans')
sc.pl.stacked_violin(adata, var_names=target_genes, groupby='kmeans')

top_genes = adata.var_names[adata.var['highly_variable']][:20]
sc.pl.dotplot(adata, var_names=top_genes, groupby='kmeans')
```

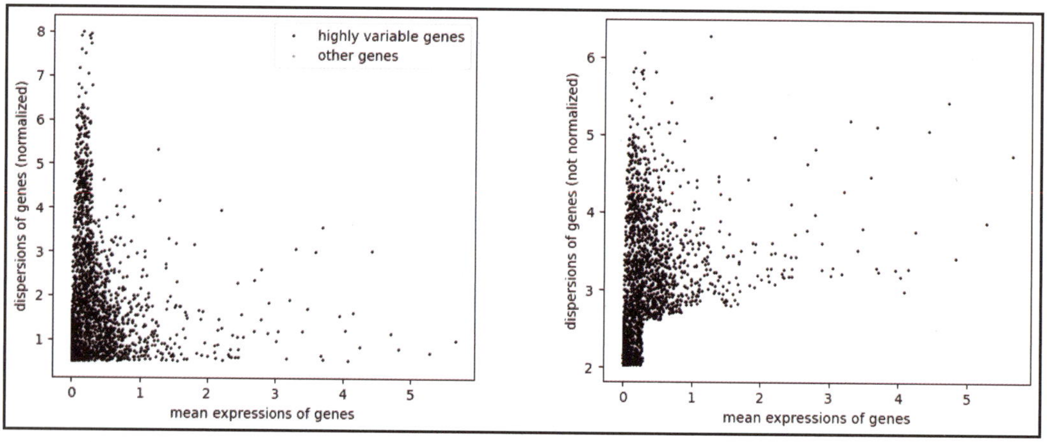

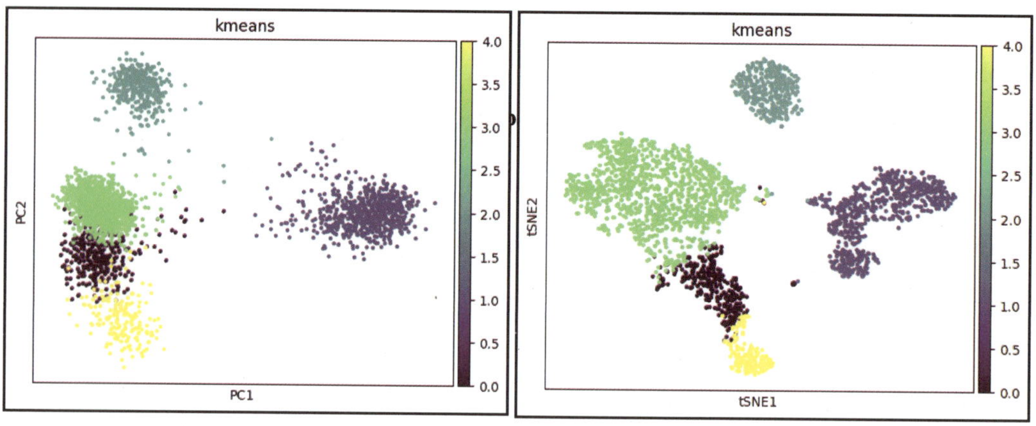

Figure 6: Dimensionality reduction of the data using PCA and tSNE

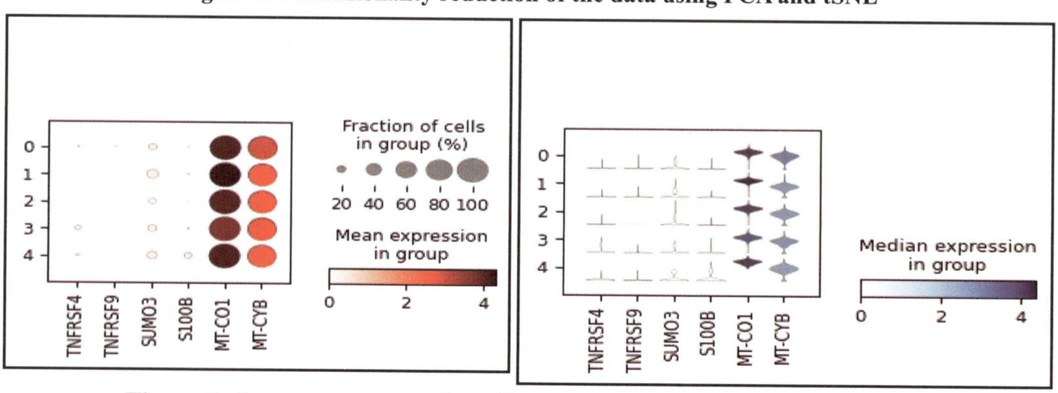

Figure 7: Showing mean and median expression in group for the genes

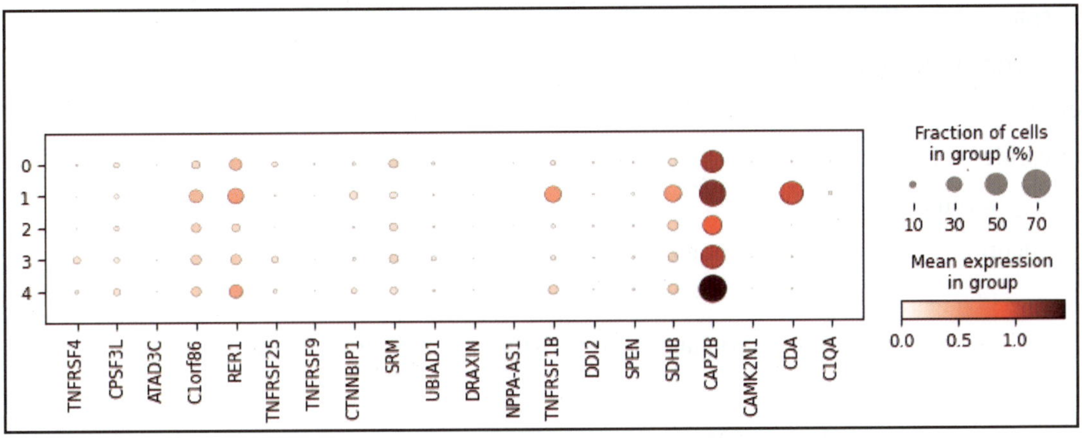

Figure 8: Showing mean expression in group for the genes

A. Proteomics

Proteomics focuses on studying proteins, their structures, and functions. AI models assist in protein structure prediction, post-translational modification detection, and large-scale proteomic analysis.

a. Protein Structure Prediction

Predicting a protein's 3D structure from its amino acid sequence is crucial for understanding its function. AlphaFold by DeepMind achieves near-experimental accuracy in structure prediction.

AI Techniques:

- **AlphaFold:** End-to-end deep learning system predicting protein structures.
- **RoseTTAFold:** Open-source alternative for protein modeling.

b. Post-Translational Modifications (PTMs)

PTMs regulate protein activity. ML models predict phosphorylation, methylation, and ubiquitination sites.

Hands-on: *Predicting Phosphorylation Sites in Kinases*

- **Objective:** Detect phosphorylation sites in kinase proteins.
- **Approach:**
  1. Use mass spectrometry datasets.
  2. Train an SVM model to classify peptides as phosphorylated/non-phosphorylated.

```
# Generate dummy training data
import random
import numpy as np

# Amino acid pool (S, T, Y are key phosphorylation sites)
amino_acids = 'ACDEFGHIKLMNPQRSTVWY'
```

```
phospho_sites = 'STY'
# Generate random peptide sequences
def generate_peptides(n, length=12, phosphorylated=False):
    peptides = []
    for _ in range(n):
        peptide = ''.join(random.choices(amino_acids, k=length))

        # Introduce phosphorylation at random positions for positive examples
        if phosphorylated:
            pos = random.choice(range(length))
            peptide = peptide[:pos] + random.choice(phospho_sites) + peptide[pos + 1:]
    peptides.append(peptide)
    return peptides

# Create datasets
phospho_peptides = generate_peptides(200, phosphorylated=True)
non_phospho_peptides = generate_peptides(200, phosphorylated=False)

# Labels (1 = phosphorylated, 0 = non-phosphorylated)
peptides = phospho_peptides + non_phospho_peptides
labels = [1] * 200 + [0] * 200
def extract_features(peptides):
    amino_acids = 'ACDEFGHIKLMNPQRSTVWY'
    features = []

    for peptide in peptides:
        counts = np.zeros(len(amino_acids))
        for i, aa in enumerate(peptide):
            # Positional weighting for S, T, Y
            weight = 1.5 if aa in 'STY' else 1
            counts[amino_acids.index(aa)] += weight
        features.append(counts / len(peptide))
    return np.array(features)
```

```
X = extract_features(peptides)

from sklearn.svm import SVC
from sklearn.model_selection import train_test_split
from sklearn.metrics import classification_report, confusion_matrix, roc_curve, auc

# Split the dataset
X_train, X_test, y_train, y_test = train_test_split(X, labels, test_size=0.2, random_state=42)

# Train SVM model
svm_model = SVC(kernel=’rbf’, C=1.0, gamma=’scale’, probability=True)
svm_model.fit(X_train, y_train)
# Evaluate model performance
y_pred = svm_model.predict(X_test)
print(“Classification Report:\n”, classification_report(y_test, y_pred))
# Confusion matrix
conf_matrix = confusion_matrix(y_test, y_pred)
print(“Confusion Matrix:\n”, conf_matrix)

import matplotlib.pyplot as plt

# Plot ROC curve
y_score = svm_model.predict_proba(X_test)[:, 1]
fpr, tpr, _ = roc_curve(y_test, y_score)
roc_auc = auc(fpr, tpr)
plt.figure()
plt.plot(fpr, tpr, color=’darkorange’, lw=2, label=f’ROC curve (AUC = {roc_auc:.2f})’)
plt.plot([0, 1], [0, 1], color=’gray’, linestyle=’--’)
plt.xlabel(‘False Positive Rate’)
plt.ylabel(‘True Positive Rate’)
plt.title(‘ROC Curve for Phosphorylation Site Detection’)
plt.legend(loc=”lower right”)
plt.show()
```

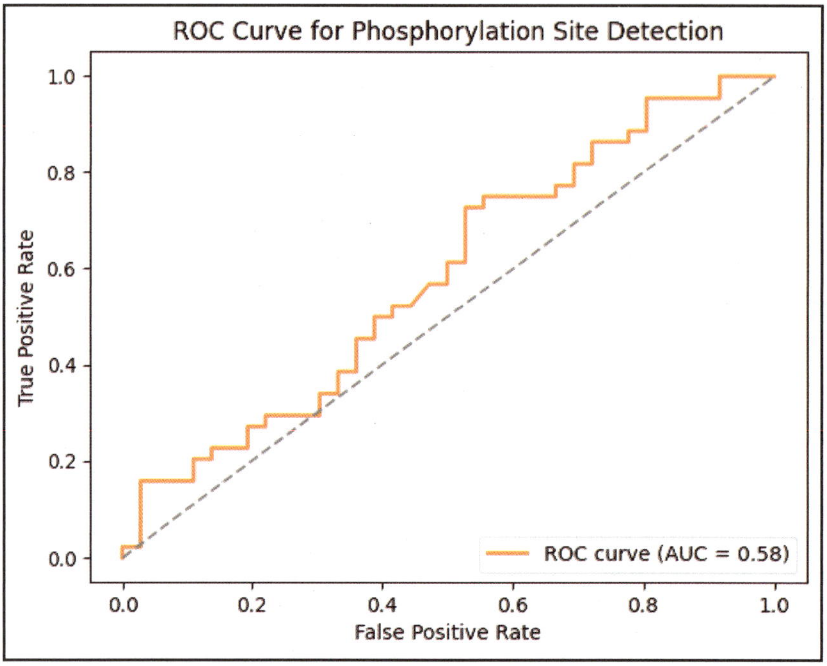

Figure 9: ROC-AUC curve

A. Biomedical Imaging

AI processes biological images for disease diagnosis and pathology.

a. Pathology

Hands-on: *AI Detection of Breast Cancer from Histopathology Slides*

- **Objective:** Automate detection of malignant cells in biopsy images.
- **Approach**
    1. Train CNNs on labeled histopathology slides.
    2. Visualize tumor regions using Grad-CAM.

```
# Import Required Libraries
import tensorflow as tf
from tensorflow import keras
from tensorflow.keras import layers, models
import numpy as np
import matplotlib.pyplot as plt
import cv2
# Download dataset (PathMNIST example)
(X_train, y_train), (X_test, y_test) = keras.datasets.mnist.load_data()
# Simulate histopathology-like images
```

```
X_train = np.expand_dims(X_train, axis=-1)  # Add channel dimension
X_test = np.expand_dims(X_test, axis=-1)
X_train = tf.image.grayscale_to_rgb(tf.convert_to_tensor(X_train))
X_test = tf.image.grayscale_to_rgb(tf.convert_to_tensor(X_test))

# Normalize
X_train = X_train / 255.0
X_test = X_test / 255.0

# Build CNN Model
model = models.Sequential([
layers.Conv2D(32, (3, 3), activation='relu', input_shape=(28, 28, 3)),
    layers.MaxPooling2D((2, 2)),
    layers.Conv2D(64, (3, 3), activation='relu'),
    layers.MaxPooling2D((2, 2)),
    layers.Conv2D(128, (3, 3), activation='relu'),
    layers.Flatten(),
    layers.Dense(64, activation='relu'),
    layers.Dense(1, activation='sigmoid')  # Binary classification])
model.compile(optimizer='adam', loss='binary_crossentropy', metrics=['accuracy'])
model.summary()
history = model.fit(X_train, y_train, epochs=10, validation_split=0.2)
```

Table 3: CNN Model Summary for Tumor Detection in Histopathology Slides

Model: "sequential"

Layer (type)	Output Shape	Param #
conv2d (Conv2D)	(None, 26, 26, 32)	896
max_pooling2d (MaxPooling2D)	(None, 13, 13, 32)	0
conv2d_1 (Conv2D)	(None, 11, 11, 64)	18,496
max_pooling2d_1 (MaxPooling2D)	(None, 5, 5, 64)	0
conv2d_2 (Conv2D)	(None, 3, 3, 128)	73,856
flatten (Flatten)	(None, 1152)	0
dense (Dense)	(None, 64)	73,792
dense_1 (Dense)	(None, 1)	65

Total params: 167,105 (652.75 KB)
Trainable params: 167,105 (652.75 KB)
Non-trainable params: 0 (0.00 B)

Conclusion

AI and ML have become indispensable tools in biological research, enabling faster, more accurate insights into complex biological systems. By understanding core algorithms, addressing ethical considerations, and staying informed about emerging trends, biologists can harness the full potential of AI to drive groundbreaking discoveries.

As AI continues to advance, its integration with biology will not only transform healthcare but also reshape the boundaries of what is possible, paving the way for a future where machines and biology work hand in hand.

Bonus content: AI and Machine Learning Applications in Biology: A Practical and Project-Based Approach

This section explores AI applications in biology with practical project ideas, hands-on approaches, and examples to engage biology students and researchers.

References

1. Pedro Larrañaga, Borja Calvo, Roberto Santana, Concha Bielza, Josu Galdiano, Iñaki Inza, José A. Lozano, Rubén Armañanzas, Guzmán Santafé, Aritz Pérez, Victor Robles, Machine learning in bioinformatics, *Briefings in Bioinformatics*, Volume 7, Issue 1, March 2006, Pages 86–112, https://doi.org/10.1093/bib/bbk007
2. Maxat Kulmanov, Fatima Zohra Smaili, Xin Gao, Robert Hoehndorf, Semantic similarity and machine learning with ontologies, *Briefings in Bioinformatics*, Volume 22, Issue 4, July 2021, bbaa199, https://doi.org/10.1093/bib/bbaa199
3. Min, Seonwoo, Byunghan Lee, and Sungroh Yoon. "Deep learning in bioinformatics." *Briefings in bioinformatics* ٨٦٩-٨٥١ :(٢٠١٧) ١٨,٥.
4. Pople, H. "Proceedings of the Fifth International Joint Conference on Artificial intelligence." (1977).
5. Zitnik, Marinka, et al. "Machine learning for integrating data in biology and medicine: Principles, practice, and opportunities." *Information Fusion* ٩١-٧١ :(٢٠١٩) ٥٠.
6. Greener JG, Kandathil SM, Moffat L, Jones DT. A guide to machine learning for biologists. *Nat Rev Mol Cell Biol*. 2022;23(1):40-55. doi:10.1038/s41580-021-00407-0
7. Tran KA, Kondrashova O, Bradley A, Williams ED, Pearson JV, Waddell N. Deep learning in cancer diagnosis, prognosis and treatment selection. *Genome Med*. 2021;13(1):152. Published 2021 Sep 27. doi:10.1186/s13073-021-00968-x
8. Hosny A, Parmar C, Quackenbush J, Schwartz LH, Aerts HJWL. Artificial intelligence in radiology. *Nat Rev Cancer*. 2018;18(8):500-510. doi:10.1038/s41568-018-0016-5
9. Kolluri S, Lin J, Liu R, Zhang Y, Zhang W. Machine Learning and Artificial Intelligence in Pharmaceutical Research and Development: a Review. *AAPS J*. 2022;24(1):19. Published 2022 Jan 4. doi:10.1208/s12248-021-00644-3
10. Shimizu H, Nakayama KI. Artificial intelligence in oncology. *Cancer Sci*. 2020;111(5):1452-1460. doi:10.1111/cas.14377

11. Becker JU, Mayerich D, Padmanabhan M, et al. Artificial intelligence and machine learning in nephropathology. *Kidney Int*. 2020;98(1):65-75. doi:10.1016/j.kint.2020.02.027

12. Tanoli Z, Vähä-Koskela M, Aittokallio T. Artificial intelligence, machine learning, and drug repurposing in cancer. *Expert Opin Drug Discov*. 2021;16(9):977-989. doi:10.1080/17460441.2021.1883585

13. Reel PS, Reel S, Pearson E, Trucco E, Jefferson E. Using machine learning approaches for multi-omics data analysis: A review. *Biotechnol Adv*. 2021;49:107739. doi:10.1016/j.biotechadv.2021.107739

14. Burns BL, Rhoads DD, Misra A. The Use of Machine Learning for Image Analysis Artificial Intelligence in Clinical Microbiology. *J Clin Microbiol*. 2023;61(9):e0233621. doi:10.1128/jcm.02336-21

10 Next-Generation Sequencing

Introduction

Next-Generation Sequencing (NGS) shines as a powerful tool of innovation with its high speed, accuracy, and scalability, NGS has transcended traditional sequencing methods, offering researchers new approaches in the field of research. NGS has opened new views of discovery, fueling breakthroughs across diverse fields, from basic research to clinical practice from the early days of Sanger sequencing to the dawn of NGS, each milestone in sequencing technology has brought us closer to the transformative capabilities we now benefit from. We'll trace this journey, exploring the triumphs and challenges that have shaped the landscape of genomic exploration. From the elucidation of the human genome to the unraveling of the complexities underlying diverse biological phenomena, NGS has emerged as the quintessential tool in the modern biologist's arsenal. With its ability to decode entire genomes, transcriptomes, and epigenomes with astonishing precision and efficiency, NGS has revolutionized our understanding of genetics, catalyzing breakthroughs across diverse domains, from basic research to clinical medicine. However, the clinical adoption of NGS brings forth a new set of challenges, particularly in bioinformatics Bioinformatics, a burgeoning field, plays a pivotal role in processing and interpreting the vast amount of data generated by NGS instruments. It encompasses the development and application of advanced computational tools to analyze high-dimensional biological data.

NGS-based bioinformatics analytics follow a structured workflow, encompassing primary, secondary, and tertiary analyses. Primary analysis involves converting raw sequencing instrument signals into nucleotide base and short-read data. Secondary analysis encompasses aligning these reads to a reference sequence or performing de novo assembly, followed by variant detection. Tertiary bioinformatics analyses contextualize the generated information by associating sample-specific genomic profiles with relevant annotation. Incorporating bioinformatics into clinical sequencing laboratories requires substantial investments in education, personnel, and infrastructure. It necessitates adaptability in processes and collaboration among stakeholders involved in testing.

Principles of Next-Generation Sequencing

Next-Generation Sequencing (NGS) technologies revolutionize genetic analysis by enabling high-throughput sequencing of DNA or RNA molecules. Unlike traditional Sanger sequencing, which processes one DNA molecule at a time, NGS technologies parallelize sequencing reactions, allowing millions of fragments to be sequenced simultaneously. This parallelization accelerates the sequencing process and reduces costs, making large-scale genomic studies feasible.

NGS Workflow Overview

The NGS workflow consists of several key steps:

Library Preparation

DNA or RNA samples are fragmented and ligated to sequencing adapters, which contain sequences necessary for attachment to the sequencing platform. This step also includes indexing samples for multiplexed sequencing, where multiple samples are pooled and sequenced together.

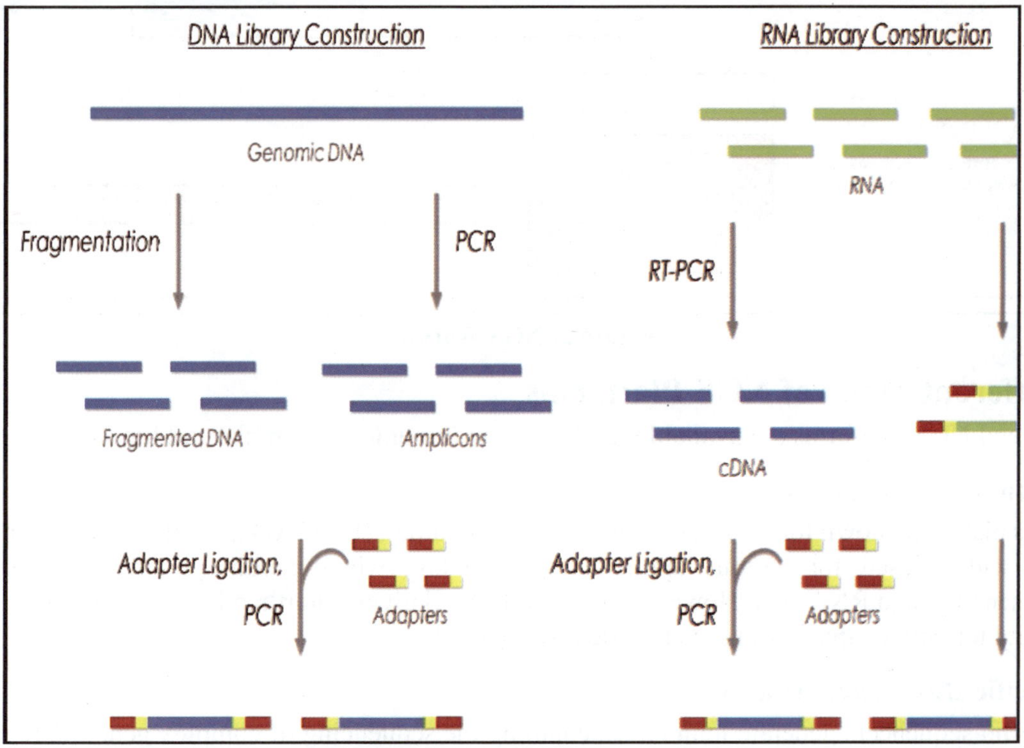

Figure1: NGS Library Construction

Sequencing Alignment

Prepared libraries are loaded onto the sequencing platform, where DNA or RNA molecules are sequenced using various technologies (such as BLAST). Each sequencing platform employs different chemistries and detection methods, but the underlying principle involves reading the sequence of bases (A, T, C, G) in each DNA or RNA fragment.

Data Analysis

Raw sequencing data is processed to generate high-quality sequence reads, which are aligned to a reference genome or assembled de novo. Variants such as single nucleotide polymorphisms (SNPs) and structural variations are identified, and biological insights are extracted through downstream analysis, such as gene expression profiling or variant annotation.

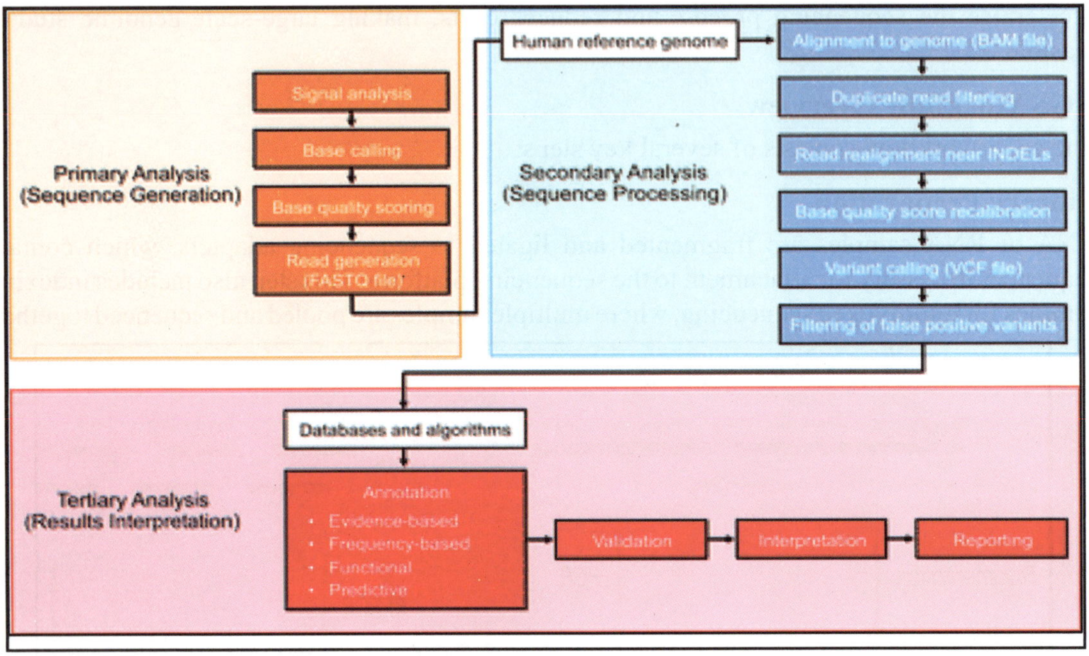

Figure 2: NGS Workflow

Different Types of NGS Platforms

Several NGS platforms are available, each of them has unique strengths and limitations:

Illumina Sequencers

This platform known for his high accuracy and fast output, (FASTAQ file) Illumina sequencers are widely used for various applications, including whole-genome sequencing, exome sequencing, and RNA-Seq. However, they produce relatively short read lengths, limiting their utility for certain applications such as de novo assembly.

Pacific Biosciences (PacBio)

PacBio sequencers generate long reads, enabling the sequencing of complex genomic regions and structural variations. They are ideal for de novo assembly and resolving repetitive sequences. However, PacBio sequencing has higher error rates compared to Illumina.

Oxford Nanopore Technologies (ONT)

ONT sequencers offer real-time sequencing with long read lengths, providing rapid insights into genomic sequences. They are portable and suitable for field applications but have higher error rates compared to Illumina and PacBio.

The choice of the NGS platform depends on the specific requirements of the study, including read length, accuracy, throughput, and cost. Illumina is favored for high-accuracy applications, while PacBio and ONT excel in generating long reads for complex genomes or real-time sequencing.

Applications of Next-Generation Sequencing

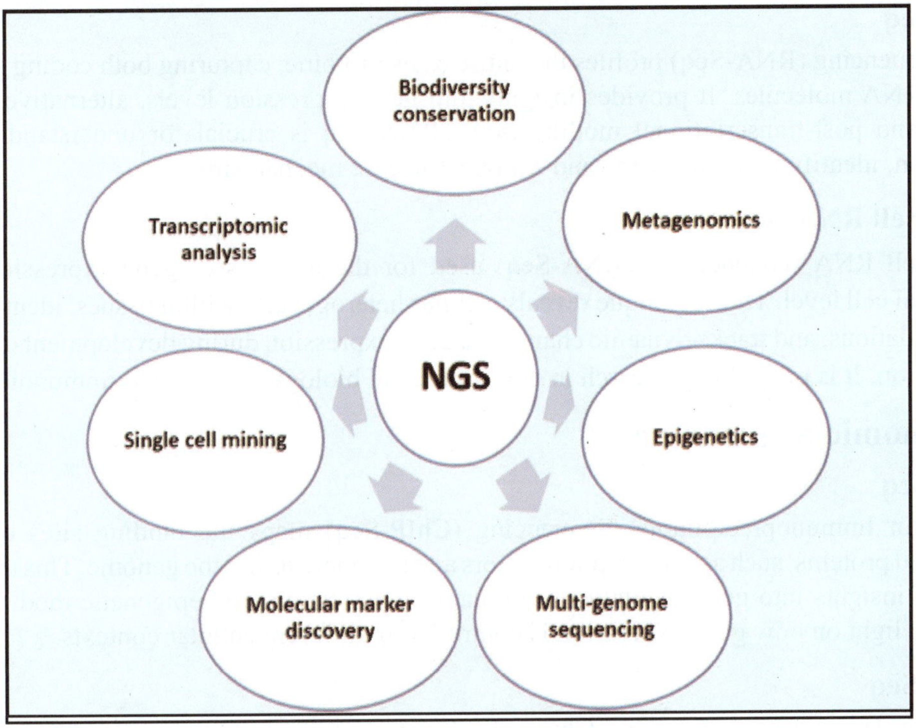

Figure 3: Application of NGS.

Genomic Sequencing

Whole Genome Sequencing (WGS)

WGS involves sequencing of the entire genome of an organism. This comprehensive approach provides a complete view of the genetic blueprint, enabling researchers to identify all genetic variations, including single nucleotide polymorphisms (SNPs), insertions, deletions, and structural variations. WGS is invaluable for studies in evolutionary biology, population genetics, and personalized medicine.

Exome Sequencing

Exome sequencing focuses on the protein-coding regions of the genome, known as exons, which represent about 1-2% of the genome but harbor 85% of known disease-related variants. This targeted approach is more cost-effective than WGS and is widely used for identifying genetic mutations linked to various diseases, especially in clinical diagnostics and research.

Targeted Sequencing

Targeted sequencing concentrates on specific genes or genomic regions of interest. This method is highly efficient and cost-effective, making it ideal for investigating specific genetic disorders, validating findings from WGS or exome sequencing, and conducting large-scale population studies where only certain regions need to be examined.

Transcriptomic Sequencing

RNA-Seq

RNA sequencing (RNA-Seq) profiles the entire transcriptome, capturing both coding and non-coding RNA molecules. It provides insights into gene expression levels, alternative splicing events, and post-transcriptional modifications. RNA-Seq is crucial for understanding gene regulation, identifying biomarkers, and studying disease mechanisms.

Single-cell RNA-Seq

Single-cell RNA sequencing (scRNA-Seq) used for the analysis of gene expression at the individual cell level. This technique reveals cellular heterogeneity within tissues, identifies rare cell populations, and tracks dynamic changes in gene expression during development or disease progression. It is pivotal for research in developmental biology, cancer, and immunology.

Epigenomic Sequencing

ChIP-Seq

Chromatin Immunoprecipitation Sequencing (ChIP-Seq) maps the binding sites of DNA-associated proteins, such as transcription factors and histones, across the genome. This technique provides insights into gene regulation, chromatin organization, and epigenetic modifications, shedding light on how gene expression is controlled in different cellular contexts.

ATAC-Seq

Assay for Transposase-Accessible Chromatin Sequencing (ATAC-Seq) identifies regions of open chromatin, indicating active regulatory elements like enhancers and promoters. This method helps in understanding chromatin accessibility and its role in gene regulation, providing a detailed view of the epigenomic landscape.

Metagenomic Sequencing

16S rRNA Sequencing:

16S ribosomal RNA sequencing targets the 16S rRNA gene, present in all bacteria and archaea, to identify and quantify microbial communities. This technique is widely used in microbial ecology, human microbiome studies, and environmental microbiology to explore microbial diversity and composition.

Metagenomic Shotgun Sequencing

Metagenomic shotgun sequencing involves sequencing all the genetic material in a sample, providing a comprehensive view of the microbial community, including bacteria, viruses, fungi, and other microorganisms. This approach allows for the analysis of microbial functions, interactions, and potential roles in health and disease.

Clinical Applications

Cancer Genomics

NGS is revolutionizing cancer research and treatment by enabling the comprehensive analysis of tumor genomes. It identifies mutations, copy number variations and other genetic alterations that drive cancer. This information guides the development of targeted therapies, monitors treatment response, and detects minimal residual disease.

Rare Disease Diagnosis

NGS is instrumental in diagnosing rare genetic disorders by uncovering pathogenic variants that might be missed by traditional methods. Exome sequencing and targeted gene panels help identify causative mutations, providing critical information for diagnosis, management, and genetic counseling.

Infectious Disease Surveillance

NGS plays a crucial role in identifying and characterizing pathogens, tracking outbreaks, and understanding the evolution of infectious agents. Metagenomic sequencing allows for the detection of known and novel pathogens directly from clinical samples, aiding in the rapid response to emerging infectious diseases.

Challenge in Next-Generation Sequencing

Data Analysis Challenges

Handling Large Volumes of Data

One of the primary challenges of NGS is managing the enormous volumes of data generated. Each sequencing run can produce terabytes of data, requiring robust computational infrastructure for storage, processing, and analysis. Efficient data management strategies and scalable computing resources are essential to handle this deluge of information.

Bioinformatics Tools and Pipelines

The complexity of NGS data necessitates sophisticated bioinformatics tools and pipelines for accurate analysis. These tools must handle various tasks, including read alignment, variant calling, and functional annotation. Developing, validating, and maintaining these pipelines is a significant challenge, as it requires expertise in both biology and computational sciences.

Accuracy and Error Rates

Sources of Errors in NGS

NGS technologies are not error-free. Common sources of errors include sequencing biases, misreads, and amplification artifacts. These errors can arise from the sequencing chemistry, the sequencing instrument, or sample preparation processes. Identifying and understanding these error sources is crucial for improving the accuracy of NGS data.

Error Correction Methods

Various error correction methods have been developed to mitigate the impact of sequencing errors. These include algorithmic approaches that correct errors based on sequence context and redundancy, as well as experimental strategies like using higher coverage to confirm variants. Advances in these methods continue to enhance the reliability of NGS results.

Ethical and Social Implications of Next-Generation Sequencing

Privacy Concerns and Data Security

The vast amount of genetic data generated by NGS raises significant privacy concerns. Genetic information is highly personal and can reveal sensitive details about an individual's health, ancestry, and predisposition to certain diseases. Ensuring the confidentiality and security of this data is paramount to prevent unauthorized access and misuse. Robust encryption methods, secure data storage solutions, and strict access controls are essential to protect individuals' genetic privacy.

Genetic Discrimination

Genetic discrimination refers to the unfair treatment of individuals based on their genetic information. This can occur in various contexts, including employment and insurance. For example, an employer might refuse to hire someone based on a genetic predisposition to a certain disease, or an insurance company might deny coverage or charge higher premiums. Legislation such as the Genetic Information Nondiscrimination Act (GINA) in the United States aims to prevent such discrimination, but concerns persist about the potential for misuse of genetic data.

Informed Consent and Genetic Counseling

Informed consent is a critical component of ethical genomic research and clinical practice. Individuals must fully understand the implications of undergoing genetic testing, including the potential risks, benefits, and limitations. Genetic counseling plays a vital role in this process, providing individuals with the information and support they need to make informed decisions about their genetic health. Counselors help interpret test results, discuss potential outcomes, and address any ethical or emotional concerns that may arise.

Equity and Access to Genomic Technologies

The benefits of NGS and other genomic technologies are not equally accessible to all populations. There are significant disparities in access to these technologies based on socioeconomic status, geographic location, and healthcare infrastructure. Ensuring equitable access to genomic testing and personalized medicine is a major ethical challenge. Efforts to address this issue include reducing the costs of sequencing, expanding healthcare coverage for genomic services, and investing in infrastructure and education in underserved communities.

Future Directions

The future of Next-Generation Sequencing (NGS) is poised for remarkable technological advancements and innovative trends. Continuous improvements in sequencing technologies are expected to further increase of accuracy in data while reducing costs. Emerging third-generation sequencing technologies, such as those from Pacific Biosciences (PacBio) and Oxford Nanopore Technologies, are leading the way with long-read capabilities that can capture complex genomic regions and structural variations more effectively. Additionally, the development of ultra-fast and portable sequencers promises to bring NGS into more diverse and decentralized settings, facilitating real-time genomic analysis in clinical and field environments. The integration of NGS with other omics technologies, such as proteomics and metabolomics, is set to revolutionize our understanding of biological systems. By combining genomic data with comprehensive analyses of proteins and metabolites, researchers can gain a more holistic view of cellular processes and disease mechanisms, enhancing our ability to unravel complex biological networks and develop novel strategies for diagnosis, treatment, and prevention.

NGS is at the forefront of personalized medicine and precision oncology, enabling tailored treatments based on an individual's genetic makeup. In oncology, NGS is used to profile the genetic landscape of tumors, identifying mutations and alterations that drive cancer progression, guiding the selection of targeted therapies and immunotherapies. Beyond cancer, personalized medicine leverages NGS to understand genetic predispositions to various diseases, optimize drug responses, and minimize adverse effects. As NGS becomes more integrated into routine clinical practice, it offers customized healthcare solutions for a wide range of conditions. However, as NGS technology advances, ethical and regulatory considerations become increasingly critical. The massive amounts of data generated by NGS pose significant privacy and security challenges, requiring stringent measures to protect individuals' genetic information. Issues of genetic discrimination and equitable access to genomic technologies must be addressed to ensure that the benefits of NGS are widely shared and do not exacerbate existing health disparities. Ethical stewardship in genomic medicine involves obtaining informed consent, providing comprehensive genetic counseling, and maintaining transparency in research and clinical applications, while regulatory frameworks evolve to keep pace with technological advancements, fostering innovation and patient trust.

Conclusion

Next-generation sequencing (NGS) has undeniably revolutionized biomedical research and clinical practice, offering unparalleled insights into the genetic underpinnings of life. Its ability to decode the human genome rapidly and accurately has transformed our understanding of biology and disease, paving the way for breakthroughs in personalized medicine, genetic research, and diagnostics. The significance of NGS in modern science cannot be overstated. From identifying genetic mutations linked to rare diseases and cancers to uncovering the complexities of microbial communities, NGS has become an indispensable tool across a multitude of disciplines. Its applications in whole-genome sequencing, transcriptomics, epigenomics, and metagenomics continue to expand, driving forward our knowledge and opening new avenues for discovery. However, as we celebrate these advancements, we must also recognize the challenges and ethical considerations that accompany the use of NGS. The vast amounts of data generated require robust analytical tools and infrastructure, and the privacy and security of genetic information must be vigilantly safeguarded. Additionally, ensuring equitable access to these transformative technologies remains a critical concern.

To fully realize the potential of NGS, ongoing research and collaboration are essential. Scientists, clinicians, bioinformaticians, and policymakers must work together to refine techniques, develop new applications, and address the ethical and social implications of genomic research. Ethical stewardship, including informed consent and genetic counseling, is paramount to maintaining public trust and ensuring that the benefits of NGS are shared widely and fairly. In closing, the journey of NGS is far from over. As we continue to push the boundaries of what is possible, let us commit to advancing the field with integrity, collaboration, and a focus on improving human health and understanding. The future of genomics holds tremendous promise, and with concerted effort and responsible innovation, we can unlock even greater discoveries and advancement.

References

1. Adnan Muhammed, U. K. (2024). NGS AND HOMOLOGY MODELING OF RB38 IN COMPLEX WITH GTP FOR NEURODEGENRATIVE DISEASE. *International Journal of Bio-Technology and Research, 14*(1), 23-38.
2. Ayman Grada, K. W. (2013). Next-Generation Sequencing: Methodology and Application. *Journal of Investigative Dermatology, 133*(8), 1-4.
3. Egalite N, G. I. (2014). Genetic counseling practice in next generation sequencing research: implications for the ethical oversight of the informed consent process. *Journal of genetic counseling, 23*, 661-670.
4. H.P.J. Buermans 1, J. d. (2014). Next generation sequencing technology: Advances and applications. *Biochimica et Biophysica Acta (BBA) - Molecular Basis of Disease, 1842*(10), 1932-1941.
5. Heydari AA, S. S. (2022). Deep learning in spatial transcriptomics: Learning from the next next-generation sequencing. *BioRxiv*, 1-29.
6. Houdayer F, P. O. (2019). Secondary findings from next generation sequencing: Psychological and ethical issues. Family and patient perspectives. *European Journal of Medical Genetics, 62*(10), 103711.
7. Hu T, C. N. (2021). Next-generation sequencing technologies: An overview. *Human Immunology, 82*(11), 801-811.
8. Jun Zhang, R. C. (2011). The impact of next-generation sequencing on genomics. *Journal of Genetics and Genomics, 38*(3), 95-109.
9. KUMARI, P. A. (2024). INTEGRATIVE ANALYSIS OF SERINE DEHYDRATASE-LIKE (SDSL) GENE IN LIVER CANCER CELLS THROUGH NGS WITH BIOPYTHON. *International Journal of Medicine and, 14*(1), 11-22.
10. Martinez-Martin N, M. D. (2019). Privacy and ethical challenges in next-generation sequencing. *Expert review of precision medicine and drug development, 4*(2), 95-104.
11. Reis-Filho, J. S. (2009). Next-generation sequencing. *Breast Cancer Research, 11*, 12.
12. Sam Behjati, P. S. (2013). What is next generation sequencing? *Archives of Disease in Childhood-Education and Practice, 98*(6), 236-238.
13. Slatko, B. E. (2018). Overview of Next-Generation Sequencing UNIT 7.31. *Current Protocols in Molecular, 122*, 59.
14. Sophia Yohe, B. T. (2017). Review of Clinical Next-Generation Sequencing. *Arch Pathol Lab Med, 141*(11), 1544-1557.
15. Uma Kumari, N. S. (2024). Validation of NGS and Multi-omics Approach for Active Site Identification CDK2-Associated Protein 1: Oral Cancer. *IJFMR, 6*(2), 1-11.
16. Xiaotu Ma, Y. S. (2019). Analysis of error profiles in deep next-generation sequencing data. *Genome Biology, 20*, 1-15.

11 Bioinformatics in Food Biotechnology

Abstract

Bioinformatics plays a pivotal role in advancing food biotechnology by leveraging computational tools for genome analysis, metabolic pathway prediction, and functional protein identification. This chapter explores the integration of bioinformatics in food production, quality enhancement, and safety, providing insights into its applications in developing genetically modified organisms (GMOs), probiotic optimization, and allergen detection.Food science is the study of food characteristics, degradation reasons, processing principles, and quality enhancement through the integration of engineering, biology, and physical sciences. Bioinformatics can be crucial to many of these procedures. The application of contemporary science would make it possible to tackle challenging issues including food deterioration, nutritional quality, and scarcity. The usage of bioinformatics techniques in food sciences is a result of numerous successful applications in other biological domains. In order for the biological data collected to be widely used in biotechnology, bioinformatics aids in the administration of biological data. The field is growing at a startling rate and is quickly becoming a vital research tool.

This discipline helps scientists explore how molecular shape determines interaction, stability, and activity. The availability of experimentally determined structures through databases such as the **Protein Data Bank (PDB)** has made it possible to analyze thousands of macromolecular conformations and derive patterns useful for structural prediction and drug discovery.

Introduction

Definition of Bioinformatics and Food Biotechnology

Bioinformatics involves the collection, analysis, and interpretation of biological data using computational tools. Food biotechnology utilizes biological systems and organisms to improve the quality, safety, and sustainability of food products.

Importance of Bioinformatics in Food Biotechnology

Identification of key genetic traits in crops and microorganisms.Enhanced food safety via pathogen detection Pevsner J. (2015)

In the field of food safety and quality, bioinformatics is becoming more and more valued. The current focus of genome sequencing projects is on food-borne pathogens and novel approaches that will aid in identifying the cause of food-borne illnesses [10]. In the future, molecular markers could be used to forecast the stress resistance of thermal preservation and identify the presence of harmful bacteria and spoiling. A technique for recognizing and detecting bacterial food pathogens has been created by bioinformatics scientists. The FDA (Food and Drug Administration) created this technology to use microarrays for the molecular characterization.

Predicting and evaluating the intended and undesirable impacts of microorganisms on food is becoming more and more dependent on bioinformatics. The impacts of microorganisms on food can be predicted using sequence-based prediction of microbial functionality and databases that combine information from genomes, systems biology, phenotypes, and "biomarkers" for functioning in certain taxa. Microbial bioinformatics can be used to maximize biomass yield, which will enhance food production. In order to do this, metabolic models and the organism's genomic sequences have been created for a variety of bacteria, including several that are relevant to food. The growth of the organism under the metabolic constraints imposed by the substrate availability in the medium can be simulated in silico using full genome-scale metabolic models and methods like flux balance analysis.

Applications of Bioinformatics in Food Biotechnology

Genomic Analysis for Crop Improvement

Genome Editing: Tools like CRISPR-Cas9 facilitate the modification of specific genes to enhance yield and disease resistance.

Data Repositories: Use of databases like NCBI, EMBL, and PlantGDB for crop genetic information (Nguyen T.H., et al. (2020)

Microbial Genomics in Fermentation

Identifying microbial strains with optimal fermentation properties.

Enhancing the production of probiotics and bioactive compounds.

Example: Lactic acid bacteria genome sequencing for improved yogurt production.

Food Safety and Allergen Detection

Computational allergenicity prediction using databases like Allergen and detection of pathogens such as E. coli and Listeria monocytogenes using metagenomic tools (Gupta R., et al. (2019)

Bioinformatics Tools and Databases in Food Biotechnology

Tools for Sequence Analysis

BLAST: Identifying sequence similarities.

Clustal Omega: Multiple sequence alignment for functional prediction.

Structural Bioinformatics

Protein Structure Prediction: Tools like SWISS-MODEL and Rosetta.

Applications in enzyme engineering for food processing.

Metabolic Pathway Analysis

Databases like KEGG and BioCyc for pathway mapping and applications in flavor enhancement and nutritional fortification(Kanehisa M., et al. (2021)

Case Studies in Food Biotechnology

Golden Rice Project

Use of bioinformatics for engineering β-carotene biosynthesis in rice.Success in combating vitamin A deficiency.

For example: Comprehensive Proteomics Analysis of Thai Jasmine Rice by Applying Next Generation Sequencing (NGS)

Rice (Oryza sativa L.), a staple food for more than half of the world's population (Shankar et al., 2016), is facing serious complications as a consequence of present changing climatic conditions. Rice holds the second position in global consumption, with India ranking next to China (Keelery, 2024). It is cultivated on nearly 160 million hectares of land worldwide, with an annual yield of around 502 million metric tons in 2022/2023 (Wailes & Chavez, 2012). India alone is estimated to produce over 135 million metric tons of rice in 2023 (Keelery, 2024). The demand to increase its production is continuously rising with the growing population size. About 2000 million metric tons of rice productivity is expected to increase by 2030 (Shankar et al., 2016).

Protein Network and Enrichment Analysis of AKR4C14

The string analysis of AKR4C14 protein showed several networks of interacting proteins out of which one protein PDC2, Pyruvate decarboxylase 2 (belongs to the TPP enzyme family) showed putative homology according to curated databases by 0.913 confidence level. This signifies the strong confidence in the accuracy of our results, indicating that the estimation method is generally reliable. Therefore, PDC2, the only characterized protein along with various other uncharacterized protein showed highest interaction with the AKR4C14 protein. The highest interaction (lowest confidence level) was observed by 0.929 confidence level with an uncharacterized protein (Figure 1).

The Interactome profiling through STRING database displayed the homology outcome with Pyruvate decarboxylase 2 (PDC2) enzyme. The network or pathways of both the enzymes were analysed connected. Both involved in metabolic pathways processing carbon compounds but with the distinct roles. PDC2 contributes in the fermentation pathway and helps regenerate NAD+ from NADH, which is essential for maintaining the glycolytic pathway under anaerobic conditions. While aldo-keto reductase enzyme reduces aldehydes and ketones to their corresponding alcohols, playing a crucial role in detoxification and metabolism of sugars, steroids, and lipids. They are interconnected through the broader metabolic pathways where pyruvate (produced from glycolysis) can enter various pathways, including those involving aldehyde or ketone intermediates that AKR might act upon. The confidence scores present the accuracy of the global networking (direct-physical and indirect-functional interactions) (Szklarczyk et al., 2019, 2023)Zenda et al. (2018), performed PPI network analysis of DAPs (Differentially Abundant Proteins) from drought stressed two varieties of maize seedling leaves (YE8112 and MO17). The several groups of interacting proteins obtained linked to amino acid metabolism, energy metabolism (NADPH production), electron transport and stress signalling, and maintaining redox homeostasis with confidence score higher than 0.5. The network analysis of DEGs (Differentially expressed genes) of Hordeum vulgare presented the networking interconnection related to metabolic processes, cellular process, localization and biological regulation (Zinati et al., 2021). Similarly, protein–protein interaction (PPI)analysis of DEGs of several plant species leaves yielded some key upregulatory and downregulatory proteins in response to drought, cold and heat stress (Balti et al., 2021).

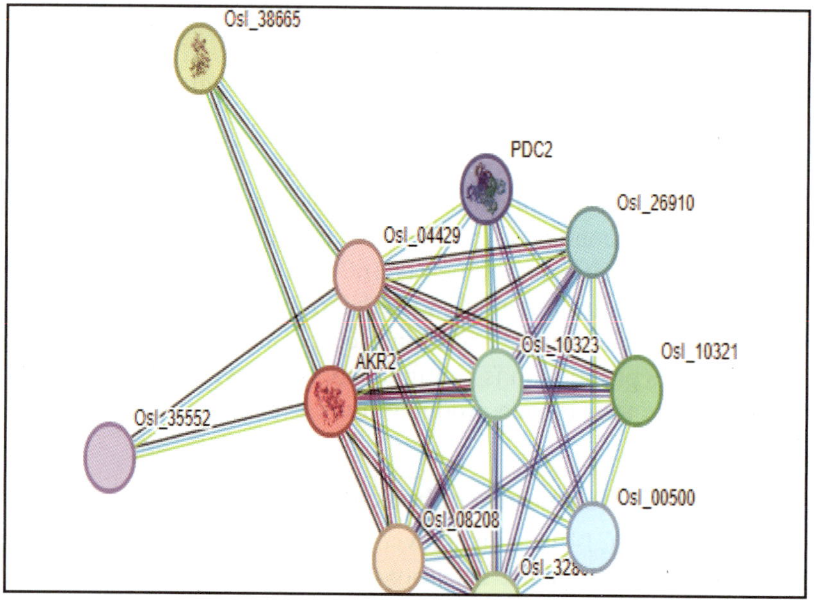

Figure 1: Protein Network and Enrichment Analysis of AKR4C14obtained from String.

Functional Analysis of AKR4C14

The functional analysis performed by Interproscan results provide the information related to several important motifs present in AKR4C14 protein. The largest region belongs to NADP-dependent oxidoreductase domain and AKR family (Figure 13). The functional analysis or gene annotation of AKR4C14 (6KBL) protein sequence is summarized in the Table 5.

This analysis reveals the key protein involved in essential biological process such as aroma production, stress resistant quality that our characteristics of Thai Jasmine Rice. The similar results were obtained for Solanum lycopersicum which displayed atleast one AKR domain in all 28 members of AKR genes (Guan et al., 2023). Shalev et al. (2018) identified terpene synthases of Thujaplicata covering 2976 identified domains across 2904 InterPro protein

Probiotic Optimization

Metagenomic studies of gut microbiota for identifying beneficial strains.Computational modeling for probiotic survival in food matrices.

Challenges and Future Directions

Challenges

Data integration from heterogeneous sources,Ethical concerns over GMO applications.High computational costs of large-scale analyses.

Future Directions

AI and machine learning integration for predictive modelling, Enhanced bioinformatics tools for personalized nutrition.

Case Studies in Food Bioinformatics

Bioinformatics has contributed significantly to solving complex problems in food biotechnology, including crop improvement, food safety, allergen detection, and microbial optimization. Below are two notable case studies illustrating its applications:

Golden Rice Project

Objective: To combat vitamin A deficiency (VAD) by engineering rice with increased β-carotene content in the endosperm.

Methodology: Genome Analysis: Bioinformatics tools identified genes responsible for β-carotene biosynthesis, such as psy (phytoene synthase) from daffodil and crtI (phytoene desaturase) from bacteria.

Gene Integration: The identified genes were introduced into the rice genome using Agrobacterium-mediated transformation.

Promoter Optimization: Computational analysis selected seed-specific promoters to ensure β-carotene expression in rice endosperm.

Tools and Databases: KEGG Pathway Database: Mapped the carotenoid biosynthesis pathway, NCBI BLAST: Verified sequence homology of inserted genes.ClustalW: Conducted multiple sequence alignment to assess conserved functional domains.

Impact: Golden rice has been developed to address malnutrition in populations dependent on rice as a staple food (Ye, X., et al. (2000) . Bioinformatics streamlined the genetic engineering process, reducing the time and cost involved, Paine, J.A., et al. (2005)

Metagenomics for Probiotic Development

To identify and optimize beneficial microbial strains for probiotic products.

Methodology: Metagenomic Analysis: Bioinformatics tools analyzed microbial diversity in fermented foods like yogurt and kimchi.16S rRNA Sequencing: Used to profile bacterial communities.

Shotgun Metagenomics: Identified functional genes related to probiotic benefits (e.g., gut adhesion, bile resistance).

Functional Prediction: Annotations from databases like MetaCyc and UniProt predicted the metabolic pathways involved.Train Engineering: Computational modeling optimized bacterial strains for survivability in food matrices and human gut environments.

Example:STRING Database: Predicted protein-protein interactions for probiotic functionality.

This approach facilitated the development of tailored probiotics targeting specific health conditions like lactose intolerance and immune health. It also improved the stability and efficacy of probiotic formulations. Fessard, A., et al. (2016)

Conclusion

These case studies illustrate the power of bioinformatics in addressing challenges in food biotechnology. From developing nutritionally enriched crops to enhancing the functionality of probiotics, bioinformatics provides tools to accelerate research and create innovative food solutions. Bioinformatics has become indispensable in food biotechnology, driving innovations in crop improvement, microbial optimization, and food safety. Despite challenges, continued advancements in computational tools promise transformative impacts on global food systems. Food science is becoming into a much more data-rich field, much like other biological sciences. Large volumes of data are produced by high throughput "omics" technologies, which necessitate data curation and analysis. In order to accomplish this, bioinformatics integration and application are appropriate. With the right software and techniques, the most recent developments in genomes, transcriptomics, proteomics, and metabolomics offer the chance to investigate and unravel the chemistry, biology, nutritional value, and micro-elements of food. Bioinformatics is a crucial field in the food sector since it allows for the effective handling of massive,volumes of data produced by numerous initiatives. Structural bioinformatics provides the foundation for understanding the molecular basis of biological function. By integrating experimental and computational data, it enables prediction, visualization, and manipulation of biomolecular structures, greatly advancing biotechnology, medicine, and genomics.

Reference

1. Pevsner J. (2015). Bioinformatics and Functional Genomics. Wiley.
2. Nguyen T.H., et al. (2020). Fermented Foods in Health Promotion and Disease Prevention. Academic Press.
3. Gupta R., et al. (2019). Bioinformatics Applications in Food Science. Springer.
4. Kanehisa M., et al. (2021). KEGG for representation and analysis of molecular networks. Nucleic Acids Research.
5. Paine, J.A., et al. (2005). Improving the nutritional value of Golden Rice through increased pro-vitamin A content. Nature Biotechnology.
6. Fessard, A., et al. (2016). Probiotic microorganisms and bioactive compounds in fermented foods: Applications and mechanisms of action. Trends in Food Science & Technology.
7. Franzosa, E.A., et al. (2018). Species-level functional profiling of metagenomes and metatranscriptomes. Nature Methods.
8. Ye, X., et al. (2000). Engineering the provitamin A (β-carotene) biosynthetic pathway into (carotenoid-free) rice endosperm. Science.
9. Paliy O., Shankar V. (2016). Application of metagenomics in studying human microbiome. Brief Bioinform.
10. Shankar, R., Bhattacharjee, A., & Jain, M. (2016). Transcriptome analysis in different rice cultivars provides novel insights into desiccation and salinity stress responses. *Scientific Reports*, *6*(1), 23719. https://doi.org/10.1038/srep23719
11. Keelery, S. (2024, March 8). *Production volume of rice India FY 2010-2023*. Statista.
12. Lesk, A. M. (2019). *Introduction to Bioinformatics* (°th ed.). Oxford University Press.
13. Szklarczyk, D., Gable, A. L., Lyon, D., Junge, A., Wyder, S., Huerta-Cepas, J., Simonovic, M., Doncheva, N. T., Morris, J. H., Bork, P., Jensen, L. J., & Mering, C. von. (2019). STRING v11: protein–protein association networks with increased coverage, supporting functional discovery in genome-wide experimental datasets. *Nucleic Acids Research*, *47*(D1), D607–D613. https://doi.org/10.1093/nar/gky1131

14. Wailes, E. J. ', & 'Chavez, E. C. '. (2012). *International Rice Baseline with Deterministic and Stochastic Projections, 2012-2021*.

15. Zinati, Z., Sazegari, S., Tahmasebi, A., & Delavari, A. (2021). A comprehensive meta-analysis to identify transcriptional signatures of abiotic stress responses in barley (Hordeum vulgare). *Cereal Research Communications*, *49*(3), 385–391. https://doi.org/10.1007/s42976-020-00107-z

16. Pevsner J. (2015). Bioinformatics and Functional Genomics. Wiley.

17. Nguyen T.H., et al. (2020). Fermented Foods in Health Promotion and Disease Prevention. Academic Press.

18. Gupta R., et al. (2019). Bioinformatics Applications in Food Science. Springer.

19. Kanehisa M., et al. (2021). KEGG for representation and analysis of molecular networks. Nucleic Acids Research.

20. Paliy O., Shankar V. (2016). Application of metagenomics in studying human microbiome.

21. Shephali Sachan, Uma Kumari, Motukuri Naveen Kumar, Muni Kumar Dokka "Drug Discovery and Homology Modeling Study of AKR4C14, an Aldo- Keto Reductase from Thai Jasmine Rice" 2024; Volume 6 , Issue -14 : Page: 8971-8984 "African Journal of Biological science" AFJBS» doi: 10.48047/AFJBS.6.14.2024.8971-8984

22. Kumari and Ashok Kumar Choudhary, "Genome sequence analysis to evaluate the performance of pairwise statistical significance of *Solanum lycopersicum* volume 2, Issue 7 (**Nov-Dec 2016)**, pp46-48, IOSR Journal (IOSR Journal of Biotechnology and Biochemistry) (IOSR-JBB)

23. Uma Kumari and Ashok Kumar Choudhary "Genome Sequence Analysis of *Solanum lycopersicum* Showing the site of Restriction Enzymes of RNA Sequence By Applying NEB Cutter As Bioinformatics Tool" .(IJBTR); NAAS rating: 3.80; VOL - 7, Issue - 3; Edition: **Jun2017** "

24. Uma Kumari "MOLECULAR ANALYSIS OF ORYZA *SATIVA* USING MEGABLAST" (IJBTR) VOL - 10, Issue - 1; Edition: **Jun2020, pp79-86**

25. Uma Kumari "Computational Analysis of Achillea Millefolium sequence to determine expectation value commonly used in Bioinformatics Database" (IJBTR), Jan **2019**.

26. Uma Kumari and Ashok Kumar Choudhary, "Computational Analysis of Solanum Lycopersicum (Tomato) Lycopersicon Esculentum Representing a Large Range of Intraspecific Variability" (IJBTR), , Vol.8, Issue 1, **2018**,

27. Uma Kumari and Ashok Kumar Choudhary,"(13th Nov 2017) Genome Sequence Analysis to Evaluate the Query Cover and E-Value Providing Genetic Statistical Significance in Lycopersicon Esculentum (IJSET) 2017, Volume 5 Issue 5 Impact Factor 4.07.

28. Uma Kumari and Ashok Kumar Choudhary , " Genome sequence analysis of *solanum lycopersicum* showing the phylogenetic relationship based on multiple sequence alignment and conserved domain proteins". (IJBTR), Vol-7, Issue-4, 2016, pp2012-2014

12 NGS With Biopython

Introduction

Next Generation Sequencing (NGS) is a powerful and readily available high-throughput tool to obtain information about all species at the genomic, transcriptional or epigenetic level. We used NGS in the field of research (e.g. metagenomics, developmental studies, and population genetics), medicine (e.g. genetic disease research, diagnosis, prognoses) and biotechnology (e.g. food safety, agriculture and agricultural improvement). Depending on specific needs, there are several sequencing platforms to choose from, each using a typical technique that results in different sequencing characteristics. Regardless of the application or even the sequencing technique chosen, the NGS workflow is characterized by three main steps: sampling, library preparation, and sequencing/analysis.

NGS has significantly encouraging genome-wide sequencing with high accuracy and determination. NGS empowers researchers to investigate complex natural questions, such as understanding the hereditary premise of maladies, unraveling developmental connections, and characterizing microbial communities. In medication, NGS plays a significant part in diagnostics, prognostics, and restorative intercessions. For occurrence, it is utilized in cancer genomics to recognize driver changes for focused on treatments, in pre-birth testing to screen for hereditary disarranges, also in irresistible illness reconnaissance to track pathogen transmission courses

The advancements in next-generation sequencing (NGS) can be traced back to the Human Genome Project, which provided the first draft sequence of the human genome using Sanger sequencing. Subsequent developments, such as the introduction of capillary electrophoresis and automated sequencing platforms, significantly increased sequencing throughput and reduced costs. The advent of NGS platforms, starting with 454 sequencing, Illumina sequencing, and others, marked a paradigm shift in DNA sequencing, making large-scale sequencing projects feasible and affordable. This progress highlights the rapid pace of technological innovation in genomics and its profound impact on biological research, medicine, and society as a whole.

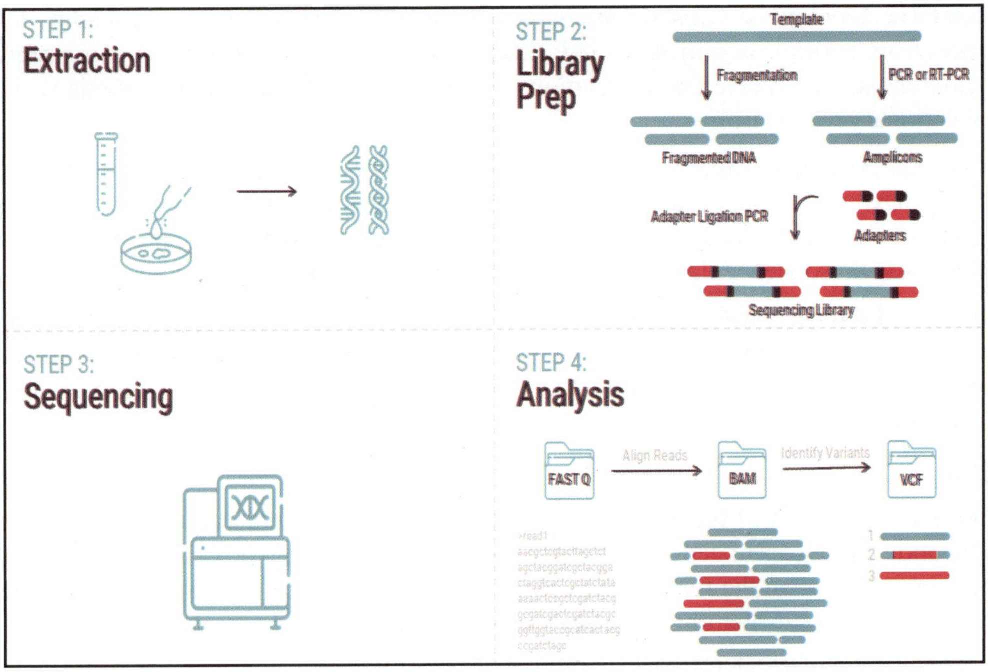

Figure 1: NGS workflow steps

NGS: Tools and Resources

Blast

BLAST is primarily used for sequence alignment. In NGS, researchers often generate short DNA or RNA sequences (reads) that need to be aligned with reference genomes or known sequences.

Functional Annotation: BLAST can be used to annotate the functional significance of DNA or protein sequences identified in NGS experiments

Chimeric Sequence Detection: NGS data may contain chimeric sequences resulting from PCR artifacts or sequencing errors. BLAST can help identify and filter out these chimeric sequences by aligning them against a database of known sequences, allowing for more accurate downstream analysis.

Metagenomics: In metagenomic studies, BLAST is used to identify and classify microbial sequences obtained from environmental samples. By aligning NGS reads against databases of microbial genomes and marker genes, BLAST facilitates the taxonomic profiling and functional characterization of microbial communities present in complex samples.

Clustal Omega

Clustal Omega is a multiple sequence alignment (MSA) tool used extensively in bioinformatics, including in Next-Generation Sequencing (NGS) applications.

Alignment of NGS Reads: In NGS experiments, researchers generate millions of short DNA or RNA sequences (reads). Clustal Omega can be used to align these reads to a reference genome or to each other.

Comparative Genomics: Clustal Omega is used in comparative genomics studies to align sequences from multiple organisms or individuals. In NGS, this can involve aligning sequences from different strains, species, or populations to identify conserved regions, genetic variations, and evolutionary relationships.

Phylogenetic Analysis: Clustal Omega alignments are often used as input for phylogenetic analysis, where evolutionary relationships among sequences are inferred. In NGS, phylogenetic analysis can be applied to study the evolutionary history and diversity of organisms, populations, or genes based on sequence data generated from high-throughput sequencing platforms.

Assembly Validation: NGS data often undergo de novo assembly to reconstruct genomes or transcriptomes from short reads. Clustal Omega can be employed to align assembled contigs or transcripts to reference sequences, helping validate the assembly and identify potential misassemblies or errors.

PyMOL

Structural Visualization of NGS-Derived Proteins: NGS techniques can be used to determine the nucleotide sequences of genes, which can then be translated into protein sequences. PyMOL can be used to visualize the three-dimensional structures of these proteins, providing insights into their folding, conformation, and potential functional sites.

Docking Studies with NGS-Derived Structures: PyMOL can be used for molecular docking studies, where the interactions between proteins and other molecules (ligands) are analyzed. NGS data may provide protein structures of interest, which can be docked with ligands to predict binding affinities and understand protein-ligand interactions relevant to drug discovery, enzyme function, or other biological processes.

Comparative Structural Analysis: PyMOL can be employed to compare the structures of proteins derived from NGS data with known structures available in databases such as the Protein Data Bank (PDB). This comparative analysis can reveal similarities and differences in protein structures, aiding in the functional annotation and characterization of NGS-derived proteins.

Integration with Structural Bioinformatics Pipelines: PyMOL can be integrated into automated pipelines for structural bioinformatics analysis, where NGS-derived data are processed, analyzed, and visualized in a streamlined manner. This integration enables researchers to efficiently visualize and interpret structural features derived from NGS experiments.

Visualization of Structural Variants: In some cases, NGS data may be used to identify structural variants in genomes, such as insertions, deletions, duplications, and inversions. PyMOL can be utilized to visualize these structural variants within the context of genomic loci or protein structures, providing insights into their potential functional implications or roles in disease.

PDB

The Protein Data Bank (PDB) is a repository of experimentally determined three-dimensional (3D) structures of proteins, nucleic acids, and complex assemblies. While PDB itself does not directly involve Next-Generation Sequencing (NGS) data, it plays several important roles in NGS research and analysis

Structural Annotation: NGS techniques often yield genetic sequences that code for proteins. The PDB serves as a valuable resource for annotating these sequences with structural information. By comparing NGS-derived protein sequences to known structures in the PDB, researchers can gain insights into the potential folding patterns, functional domains, and structural features of the proteins encoded by the sequenced genes.

Molecular Docking and Virtual Screening: NGS research often involves identifying potential drug targets or understanding protein-ligand interactions. The PDB provides a wealth of structural data on proteins bound to small molecules, peptides, and other ligands.

Comparative Structural Analysis: The PDB enables comparative structural analysis of proteins encoded by NGS-derived sequences. Researchers can compare the structures of homologous proteins, identify conserved structural motifs, and explore structural variations associated with genetic polymorphisms or disease-associated mutations identified through NGS studies.

Figure 2: PDB Webpage

COBALT

COBALT (Constraint-based Multiple Protein Alignment Tool) is a bioinformatics tool developed by the National Center for Biotechnology Information (NCBI) primarily for multiple sequence alignment (MSA) of proteins. While it's not directly related to Next-Generation Sequencing (NGS) data analysis.

Comparative Genomics: COBALT can align protein sequences obtained from NGS data with sequences from related organisms or species. This comparative analysis allows researchers to identify conserved regions, evolutionary relationships, and functional similarities among proteins across different genomes.

Variant Analysis: NGS data may reveal genetic variations, such as single nucleotide polymorphisms (SNPs) or insertions/deletions (indels), within protein-coding regions of the genome. COBALT can align protein sequences harboring these variations to identify their locations relative to conserved protein domains or functional motifs

Phylogenetic Analysis: COBALT-generated alignments can be used as input for phylogenetic analysis to reconstruct evolutionary relationships among protein sequences derived from NGS data. By building phylogenetic trees based on COBALT alignments, researchers can infer the evolutionary history and divergence patterns of protein families or functional groups, providing insights into the molecular evolution of organisms or biological processes.

Structural Bioinformatics: COBALT alignments can aid in the prediction and analysis of protein structures derived from NGS data. By identifying conserved regions and structurally important residues within aligned protein sequences, COBALT helps guide comparative modeling and structural prediction efforts.

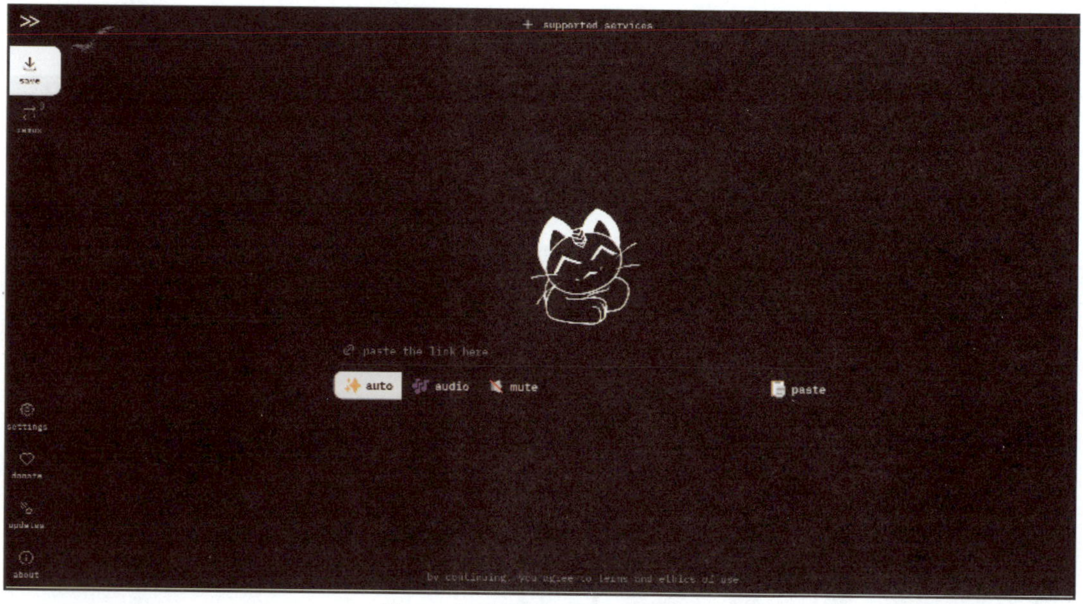

Figure 3: Cobalt tool webpage

NGS with python

NGS technologies have revolutionized genomic research by enabling the rapid sequencing of entire genomes. Bio-python offers many functions to handle large genomic datasets, perform sequence alignments, analyze genetic variations, and annotate genomic features. NGS has become more important in clinical settings for diagnosing genetic disorders, predicting disease risks, guiding personalized medicine, assists in analyzing clinical NGS data, interpreting genetic variants, and predicting the functional impact of mutations. NGS data are valuable for studying evolutionary relationships, population genetics, and adaptive evolution, provides tools for phylogenetic analysis, molecular evolution studies, and comparative genomics, facilitating research in evolutionary biology. Bio-python serves as an educational resource for learning bioinformatics and computational biology, as it offers accessible and well-documented functionalities for handling biological data and performing common analysis tasks using Python programming language.

Importing Libraries

First we have to import necessary libraries such as Bio-python for sequence analysis and other standard libraries for file handling and data manipulation.

```python
from Bio import SeqIO
from Bio.SeqIO.QualityIO import FastqGeneralIterator
```

Loading Data

After import necessary files next, we load the raw sequencing data in FASTQ format. Bio-python provides utilities to efficiently parse FASTQ files.

```python
input_file = "raw_data.fastq"
```

Quality Filtering and Trimming

Now we trim low-quality bases and filter out reads that do not meet certain quality thresholds for better accuracy. Bio-python offers functions to filter sequences based on quality scores.

```python
def filter_sequences(input_file, output_file, min_qual=20, min_length=50):
    with open(output_file, "w") as out_handle:
        for title, seq, qual in FastqGeneralIterator(open(input_file)):
            if min(qual) >= min_qual and len(seq) >= min_length:
                out_handle.write("@%s\n%s\n+\n%s\n" % (title, seq, qual))

filter_sequences(input_file, "filtered_data.fastq")
```

Then we are ready to use bio-python coding for our sample.

Application of Next Generation Sequencing With Biopython in Bioinformatics

Bio-python is a powerful and versatile toolkit for manipulating and analyzing next-generation sequencing (NGS) data. It provides modules to handle various NGS file formats, perform sequence alignments, and offer utilities for data analysis.

NGS Data Manipulation and Analysis

Bio python give us modules like SeqIO, AlignIO and SeqUtils that can handle various NGS file formats. These modules facilitate sequence alignments and provide utilities for data analysis.

It supports many files format like FASTA, GENEBANK, JSON AND GTF which we commonly used in Bioinformatics.

Quality Control (QC)

Bio python is used for quality control to ensure reliable and accurate NGS data analysis. It helps identify and mitigate potential issues such as sequencing errors, adapter contamination, and low-quality reads. SeqIO and SeqUtils modules are used for NGS data manipulation and QC analysis. Common QC metrics include read quality scores (Phred scores), read length distribution, GC content, adapter contamination, duplicated reads, and sequence complexity.

Interfacing with External NGS Tools

Bio-python can seamlessly integrate with external NGS tools, allowing utilization of their features. It is compatible with tools such as:

BWA (Burrows-Wheeler Aligner): For read alignment against a reference genome.

Bowtie: A fast and memory-efficient read alignment tool.

GATK (Genome Analysis Toolkit): A comprehensive suite for variant calling, genotyping, and variant analysis.

MEME Suite : A collection of tools for motif discovery and analysis in DNA, RNA, and protein sequences.

BED Tools: A set of utilities for manipulating genomic intervals and working with BED file formats.

Sequence Alignment

Bio python provides modules for performing pairwise and multiple sequence alignments using algorithms like Needleman-Wunsch, Smith-Waterman, and ClustalW.

Phylogenetic Analysis

Bio python supports phylogenetic analysis, including tree construction, manipulation, and visualization.

Population Genetics

Bio-python have modules which used in population genetics to analyse allele frequency and Hardy Weinberg equation.

Biological Databases

Bio-python provide us accesses of various biological database like NCBI,UNIport for searching and retrieve more accurate database.

Example: NGS Data Quality Control with Bio python

```python
from Bio import SeqIO
from Bio.SeqUtils import GC
# Read FASTQ file
sequences = SeqIO.parse("sample.fastq", "fastq")
# Perform quality control
for sequence in sequences:
```

```
    if sequence.letter_annotations["phred_quality"] >= 20 and GC(sequence.seq) >= 50:
        print(sequence.id, "passed quality control")
    else:
        print(sequence.id, "failed quality control")
```

This script reads a FASTQ file, checks the Phred score and GC content of each sequence, and determines whether each sequence passes the quality control criteria.

Limitation of Biopython In Next Generation Sequencing

Bio-python is a valuable tool for NGS data analysis but it has certain limitations:

External Tools

NGS data analysis mostly requires specific tools for work like read alignment, variant calling, and motif finding, which are not directly available in Bio-python. Bio-python is compatible with tools such as BWA, Bowtie, GATK, MEME Suite, and BED Tools which allowing integration and utilization of their features.

Data Volume

NGS generates huge amounts of data, requiring efficient storage and processing. Bio-python may not be that efficient that other specific tools for handling extremely large datasets.

Bioinformatics Expertise

Proper NGS data analysis requires good knowledge of bioinformatics tools and algorithms. Bio-python is a toolkit that requires some programming knowledge and well-known with bioinformatics concepts.

Data Interpretation

Extracting meaningful biological insights from NGS data is complex and requires advanced analysis techniques. Bio-python provides tools for data manipulation and analysis, but interpreting the results requires biological knowledge and expertise.

Processing Speed

Bio-python uses text files and processing this highly text-based file formats can be time-consuming.

Constantly Evolving Field

NGS tools are updated often, so new tools can be regularly applied when more NGS data is made available. Collecting all the necessary information from publications, analysis protocols, and forums can be a time-intensive process, especially since it's often challenging to judge the up-to-detente sand relevance of the provided information.

Future Prospectus of Bio-Python With Next Generation Sequencing

The future of Bio-python with next-generation sequencing (NGS) involves several key trends and advancements:

Faster and Cheaper Sequencing

NGS platforms are becoming faster and more affordable, which will lead to more frequent sequencing of genomes across a wider range of applications.

More Accurate Sequencing

NGS technologies are also becoming more accurate, which is crucial for clinical diagnostics where even small errors can have serious consequences.

Longer Reads

NGS technologies are continuously evolving to read much longer stretches of DNA, modify applications such as genome assembly and structural variation analysis. Long-read sequencing can generate reads up to 25–30 kb, whereas short-read sequencing can generate reads around 600–700 bp.

Single-Cell Sequencing

NGS is mostly used to sequence the DNA of individual cells, which is a important tool for understanding cell-to-cell heterogeneity in biological processes.

Epigenomics

NGS is being applied to study the epigenome, which involves the chemical modifications that control gene expression, offering new insights into biology and disease.

Personalized Medicine

NGS offers new opportunities for personalized precision medicine.

NGS and Bioinformatics Workflows

NGS incorporates with bioinformatics is a impressive tool for the detection, identification, and analysis of human pathogens.

Third-Generation Sequencing

After Next Generation Sequencing, Third-generation sequencing technologies like nanopore sequencing (Oxford Nanopore) and single-molecule real-time (SMRT) sequencing (PacBio), offer long read lengths and real-time sequencing capabilities.

Conclusion

Utilizing Bio-python for NGS analysis has significantly advanced researchers' ability to decipher the intricacies of genomic and transcriptomic data, offering substantial insights into biological systems and disease mechanisms. Bio-python's robust tools and algorithms allow scientists to conduct thorough analyses, including quality control, read alignment, variant calling, and differential expression analysis. These discoveries have critical implications for biomedical research, personalized medicine, and the broader comprehension of molecular biology. Looking ahead, the continued integration of omics data, sophisticated computational techniques, and new sequencing technologies will foster ongoing innovation and breakthroughs in genomics.

References

1. https://upstemacademy.com/courses/biopython-fundamentals/lesson/introduction-to-ngs-data-and-file-formats/
2. https://upstemacademy.com/courses/biopython-fundamentals/lesson/introduction-to-biopython-compatible-ngs-tools/
3. https://goldbio.com/articles/article/Four-Steps-for-NGS-Data-Analyses
4. https://www.packtpub.com/en-us/learning/how-to-tutorials/processing-next-generation-sequencing-datasets-using-python/
5. https://www.ecseq.com/support/ngs/getting-started-with-ngs-data-analysis-overview
6. https://unicornplatform.com/domain-not-supported/?domain=dataheadhunters.com
7. https://omicstutorials.com/mastering-bioinformatics-with-biopython-a-comprehensive-guide/

13 Structural Bioinformatics

Introduction

In Bioinformatics structural biology focuses on the computational analysis,molecular modelling the analysis and interpretation of the 3D structures of biological macromolecules, such as proteins, nucleic acids, and molecular complexes (Lesk, A. M. (2004) By integrating computational techniques with structural biology data, it provides crucial insights into molecular function, dynamics, and interactions. Applications include drug discovery, protein engineering, and understanding the structural basis of diseases. Key methodologies include homology modeling, molecular docking, and molecular dynamics simulations, supported by tools like PyMOL, MODELLER, and AutoDock. Structural bioinformatics also leverages databases like the Protein Data Bank (PDB) to analyze biomolecular structures. Despite challenges like data gaps and computational complexity, advancements in machine learning and cryo-electron microscopy are driving the field forward. This discipline is pivotal in bridging sequence information and biological function, offering transformative solutions in precision medicine and biotechnology. Structural bioinformatics is a specialized branch of bioinformatics that focuses on the analysis, prediction, and visualization of the three-dimensional (3D) structures of biological macromolecules such as proteins, nucleic acids, and complexes. It bridges computational biology and structural biology, providing tools and methodologies to understand the relationship between molecular structure and biological function (Lesk, 2019).

Key focus areas include:

Predicting molecular structures.

Modeling molecular interactions.

Analyzing structural changes caused by mutations.

Importance of Structural Bioinformatics

Structural bioinformatics plays a pivotal role in modern biology and medicine, offering:

Functional Insights: The 3D structure of biomolecules provides detailed information about active sites, binding pockets, and interaction domains, which are essential for understanding molecular mechanisms.

Drug Discovery: It helps identify and optimize molecules that interact with specific biomolecular targets.

Disease Understanding: Structural bioinformatics links genetic mutations to structural changes, aiding in the understanding of diseases such as cancer and Alzheimer's.

Fundamental Concepts in Structural Bioinformatics

Protein Structure Levels

Proteins exhibit four hierarchical levels of structure: primary, secondary, tertiary, and quaternary.

- Primary structure refers to the linear sequence of amino acids.
- Secondary structure includes local conformations such as α-helices and β-sheets, stabilized by hydrogen bonds.
- Tertiary structure represents the overall 3D folding of a single polypeptide chain, while
- Quaternary structure describes the arrangement of multiple subunits within a protein complex (Nelson & Cox, 2021).

Key Methods in Structural Bioinformatics

Experimental Techniques

X-ray Crystallography: Produces high-resolution 3D structures of biomolecules.

Nuclear Magnetic Resonance (NMR): Determines structures in solution, capturing dynamics.

Cryo-Electron Microscopy (Cryo-EM): Resolves large molecular complexes.

Computational Approaches in Structural Bioinformatics

Homology Modeling

Homology modeling predicts the 3D structure of a protein using a known template with similar sequence identity. It assumes that structurally similar sequences share common folds and functions (Šali & Blundell, 1993). Tools like SWISS-MODEL and MODELLER are widely used for this approach.

Molecular Docking

Molecular docking simulates the interaction between a receptor (protein) and a ligand (drug or small molecule). It helps predict binding affinity and orientation within the active site, aiding drug discovery and virtual screening (Morris et al., 2009).

Molecular Dynamics (MD) Simulation

MD simulations analyze atomic motion over time under defined physical conditions. They provide insights into protein folding, conformational flexibility, and stability under different environments (Karplus & McCammon, 2002).

Structure Validation

After computational modeling, structural validation is crucial to ensure accuracy. Tools such as PROCHECK, Verify3D, and Ramachandran plots assess stereochemical quality and identify modeling errors (Laskowski et al., 1993).

Challenges in Structural Bioinformatics

Data Gaps: Limited availability of experimentally determined structures for many proteins.

Complexity: Biomolecules are large, flexible, and dynamic, complicating structural predictions.

Computational Demands: Simulations and analyses require significant computational resources

Importance of Structural Bioinformatics

Understanding Function: The 3D structure of biomolecules often determines their function, enabling the study of active sites, binding domains, and interaction interfaces.

Drug Design: Structural bioinformatics aids in identifying and designing small molecules that bind to specific biomolecular targets.

Molecular Mechanisms: It elucidates mechanisms like enzyme catalysis, protein folding, and molecular signaling.

Structural Variants: Analysis of structural changes, such as mutations, helps link molecular alterations to diseases.

Key Concepts and Methods

Structural Representation Of Biomolecules

Primary Structure: Linear sequence of amino acids or nucleotides.

Secondary Structure: Localized folding into helices (e.g., α-helix) or sheets (e.g., β-sheet).

Tertiary Structure: The overall 3D conformation of a single molecule.

Quaternary Structure: Complexes formed by multiple biomolecules.

Techniques for Structural Determination

X-ray Crystallography: Provides atomic resolution for proteins and DNA.

Nuclear Magnetic Resonance (NMR): Captures dynamic structures in solution.

Cryo-Electron Microscopy (Cryo-EM): Resolves large molecular complexes.

Structural Databases

Protein Data Bank (PDB): Repository for 3D structures of proteins and nucleic acids.

CATH: Classification systems for protein structural domains.

DSSP: Database for secondary structure assignments.

Homology Modeling

Predicts the 3D structure of a protein based on its similarity to a known structure.

Molecular Dynamics (MD) Simulations

Simulates the movement of molecules over time to study dynamics and stability.

Tools: GROMACS, AMBER.

Docking Studies

Predicts the interaction between two molecules (e.g., protein-ligand or protein-protein).

Tools: AutoDock, HADDOCK.

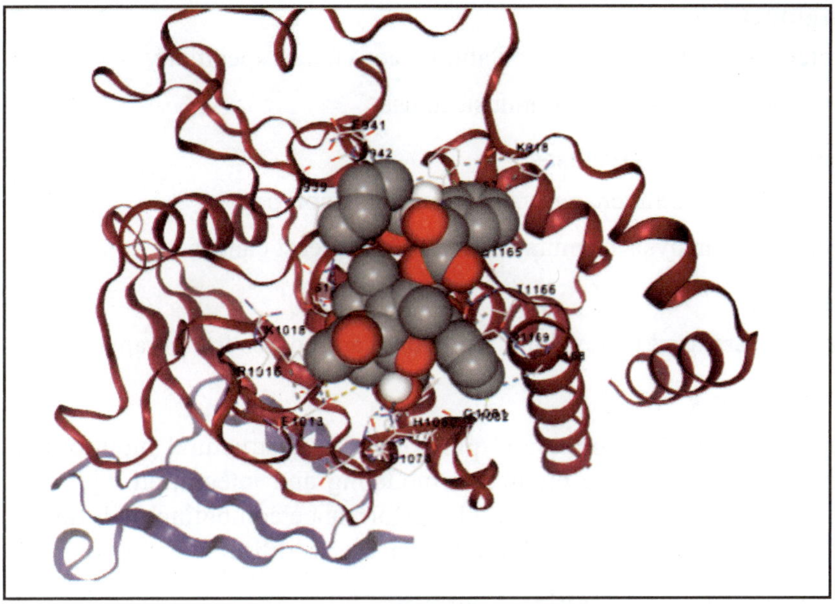

Figure 11: Docking with Taxol C

Table 1: Docking score with Taxol C

CurPocket ID	Vina Score	Cavity Volume (Å³)	Center (x, y, z)	Docking Size (x, y, z)
C4	-6.9	281	-19, 34, -23	26, 26, 26
C3	-6.4	330	-12, 9, -20	26, 26, 26
C2	-5.4	641	-31, 25, 0	26, 26, 26
C1	-5.1	817	-13, 17, -7	26, 26, 26
C5	-5.1	239	10, 31, -12	26, 26, 26

The table summarizes the docking characteristics of five predicted binding pockets. Among them, C4 exhibits the most favorable Vina score of -6.9, indicating the strongest binding affinity. It is followed by C3 with a score of -6.4. Although C1 has the largest cavity volume (817 Å³), its Vina score is lower, suggesting weaker binding potential despite the spacious pocket.

Applications of Structural Bioinformatics

Drug Discovery and Development

Identifies drug targets and optimizes lead compounds using molecular docking and virtual screening.

Example: Design of kinase inhibitors for cancer treatment.

Protein Engineering

Modifies protein structures to improve stability, activity, or specificity.

Example: Engineering enzymes for industrial use.

Understanding Genetic Diseases

Links mutations to structural changes and disease phenotypes.

Example: Structural analysis of mutated BRCA1 in breast cancer.

Antibody Design

Structural insights are used to develop monoclonal antibodies for therapy.

Future Directions

With the rapid growth of genomic and proteomic data, structural bioinformatics continues to evolve toward more accurate predictive modeling and integration with systems biology. The combination of AI, cryo-EM, and high-performance computing will further enhance the precision of structural predictions and drug discovery workflows

Integration with AI and Machine Learning

Tools like AlphaFold revolutionize structure prediction with high accuracy.

Machine learning models enhance docking and binding affinity predictions.

Structural Systems Biology

Combines structural data with network biology to study interactions at a systems level.

Cryo-EM Advancements:Improved resolution in Cryo-EM expands its application to challenging molecular systems.

Conclusion

Structural bioinformatics bridges the gap between sequence data and functional insights by focusing on the 3D structures of biomolecules. Its applications in drug discovery, disease understanding, and protein engineering make it a cornerstone of modern bioinformatics. Continuous advancements in computational tools and structural biology techniques are set to further enhance its capabilities in solving complex biological problems.

Reference

1. Lesk, A. M. (2004). Introduction to Bioinformatics. Oxford University Press.
2. Bourne, P. E., & Gu, J. (2010). Structural Bioinformatics. Wiley-Blackwell.
3. Mount, D. W. (2004). Bioinformatics: Sequence and Genome Analysis. Cold Spring Harbor Laboratory Press.
4. Pettersen, E. F., et al. (2004). UCSF Chimera—A visualization system for exploratory research and analysis. Journal of Computational Chemistry, 25(13), 1605–1612.
5. Sali, A., & Blundell, T. L. (1993). Comparative protein modelling by satisfaction of spatial restraints. Journal of Molecular Biology, 234(3), 779–815.

6. Berman, H. M., Westbrook, J., Feng, Z., Gilliland, G., Bhat, T. N., Weissig, H., ... & Bourne, P. E. (2000). The Protein Data Bank. *Nucleic Acids Research, 28*(1), 235–242.

7. Chothia, C., & Lesk, A. M. (1986). The relation between the divergence of sequence and structure in proteins. *EMBO Journal, 5*(4), 823–826.

8. Jumper, J., Evans, R., Pritzel, A., Green, T., Figurnov, M., Ronneberger, O., ... & Hassabis, D. (2021). Highly accurate protein structure prediction with AlphaFold. *Nature, 596*(7873), 583–589.

9. Karplus, M., & McCammon, J. A. (2002). Molecular dynamics simulations of biomolecules. *Nature Structural Biology, 9*(9), 646–652.

10. Laskowski, R. A. (2009). Structural bioinformatics: methods and applications. *Bioinformatics, 25*(17), 2223–2224.

11. Laskowski, R. A., MacArthur, M. W., Moss, D. S., & Thornton, J. M. (1993). PROCHECK: a program to check the stereochemical quality of protein structures. *Journal of Applied Crystallography, 26*(2), 283–291.

12. Morris, G. M., Huey, R., Lindstrom, W., Sanner, M. F., Belew, R. K., Goodsell, D. S., & Olson, A. J. (2009). AutoDock4 and AutoDockTools4: automated docking with selective receptor flexibility. *Journal of Computational Chemistry, 30*(16), 2785–2791.

13. Murzin, A. G., Brenner, S. E., Hubbard, T., & Chothia, C. (1995). SCOP: a structural classification of proteins database for the investigation of sequences and structures. *Journal of Molecular Biology, 247*(4), 536–540.

14. Nelson, D. L., & Cox, M. M. (2021). *Lehninger Principles of Biochemistry* (^th ed.). W. H. Freeman.

15. Orengo, C. A., Michie, A. D., Jones, S., Jones, D. T., Swindells, M. B., & Thornton, J. M. (1997). CATH–a hierarchic classification of protein domain structures. *Structure, 5*(8), 1093–1108.

16. Wang, Y., et al. (2002). MMDB: annotating protein sequences with structural information. *Nucleic Acids Research, 30*(1), 249–252.

14 Application of Bioinformatics in COVID19

Introduction

Bioinformatics is rapidly growing and fascinating field that combine biology, computer science and statistics to analyze and interpret essentially important large biological data mainly DNA sequences and Protein structure. Bioinformatics have huge range of techniques and tools like algorithms, databases and software for produce meaningful results of datasets. The term Bioinformatics first coined by Dutch biologist Paulien Hogeweg and Ben Hesper in 1970, they described it as the information process in biotic system. The Human Genome Project in 1990 was a great milestone in bioinformatics history.it was an integral strategy to code the human gene map and design the new computer programme for storing, processing and interpreting the huge load of genetic information. Bioinformatics have many key components like data collection, data analysis, data storage and visualization. There are many bioinformatics tools that served as the main tools for the study of how genetic variation, gene expression and disease work also there are many functional structural tools for analysing protein structure that reproducing molecular integration. Nowadays Data interpretation in bioinformatics is rapidly evolving to the advancement of sequence technology, machine learning and data integration methods. The field of bioinformatics evolved in multidisciplinary area involving in the field of personalized healthcare, micro biome examination, engineering with the use of computers, Artificial intelligence and machine education on biological data and do this with the help of biologist, computer scientist, statisticians and mathematicians. [1,2]

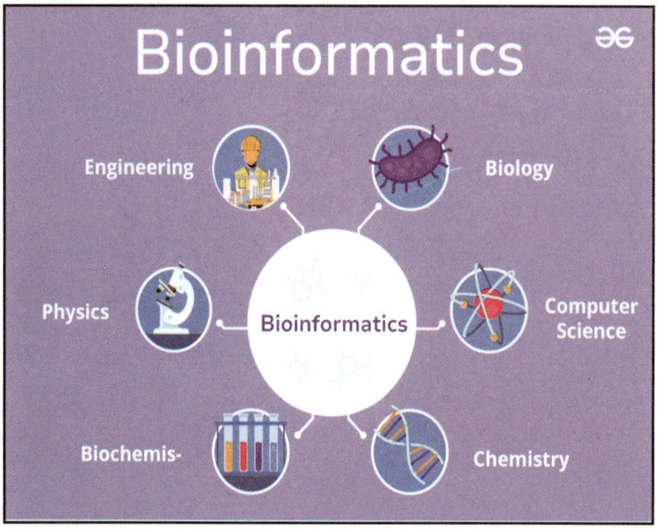

(Figure taken from GeeksforGeeks | A computer science portal for geeks for only understanding purpose)

Overview of SARS CoV-2 COVID-19 disease

The COVID-19 widespread globally, caused by the Sever Acute Coronavirus syndrome 2 (SARS CoV-2) turn up in December 2019 in Wuhan, China. The WHO (world health organization) declare it the public health emergency of international concern on January 30, 2020 and later classified it as a pandemic on March 11, 2020. The pandemic of covid 19 has also highlighted the interconnection of global health, economy and society, revealing vulnerabilities in our systems and the need for robust public health infrastructure. SARS CoV-2 related to SARS-CoV 2 and Middle East respiratory syndrome coronavirus (MERS-CoV). SARS-CoV-2 is believed to cross species to initiate primary human infection and now spread by human to human transmission. Although the case fatality rate of COVID-19 (estimated at 2%–3%) is lower than those of SARS (approximately 10%) and MERS (approximately 40%), the pandemic associated with COVID-19 has been far more severe.

Virology and Immunology of Coronavirus

Structure of Coronavirus

Coronavirus are indeed fascinating and complex structure. There are four main subgroup – alpha, beta, gamma and delta. Alpha group includes human pathogen CoV- 229E and CoV-HKU1 and Beta group includes human pathogens CoV-OC43, SARS CoV, MERS-CoV, and SARS- CoV-2. SARS-CoV-2 classified into AB coronavirus. The amino acid sequence with in the seven conserved domains of the genomic open reading frame 1ab (ORF 1ab) are 96.6% identical to the original SARS-CoV. Coronavirus typically round or multi shaped. Tis size measures 120-160nm in diameter. It includes a petal shaped projection like structure called spike (s) protein and it is a common feature of coronavirus, it mediates virus attachment and membrane fusion during infection. (King et al., 2012). Other than S protein structure coronavirus genome includes three additional structural proteins like, the Membrane (M) protein, the Envelope (E) protein and the Nucleocapsid (N) protein.

The membrane protein consist of 218-263 amino acid with an N- terminus modified by an O or N glycan and a hydrophilic C-terminal tail. The Envelope protein consist of 74-109 amino acids and actively involved in virulence with about 20 copies per virion. The Nucleocapsid protein consist of 349-470 amino acid and Ti is a RNA bound phosphorylated protein that participate in the folding of genomic RNA into the nucleocapsid protein. [3, 4]

Genomic Speciality of SARS-CoV-2

The SARS-CoV-2 virion measures 60-140nm in diameter and it is a positive sense single stranded 29891 base pair RNA genome. Genome sequence alignment of SARS CoV-2 is 79% identical to SARS CoV and 93.1% similar to the RaTG12 virus which is isolated from a bat in Yunnan Province, China, and that remarkable identity suggests that SARS-CoV-2 may have originated from a virus endemic in the bat species. The SARS-CoV-2 genome contains nearly thirty open reading frame (ORFs) and two novel insertion. Comparative analysis show that the sequences of ORF6, ORF8, and the S gene exhibit a relatively low degree of sequence conservation among coronvirus1. Although the genome of bat-SL-CoVZC45, bat-SL-CoVZXC21 and RaTG13 viruses are quite similar to SARS –Cov-2 with respect to ORF8. ORF8plays role in modulating the host immune system including down regulating of MHC class 1 molecules that helps the virus evade the immune system.[3,4]

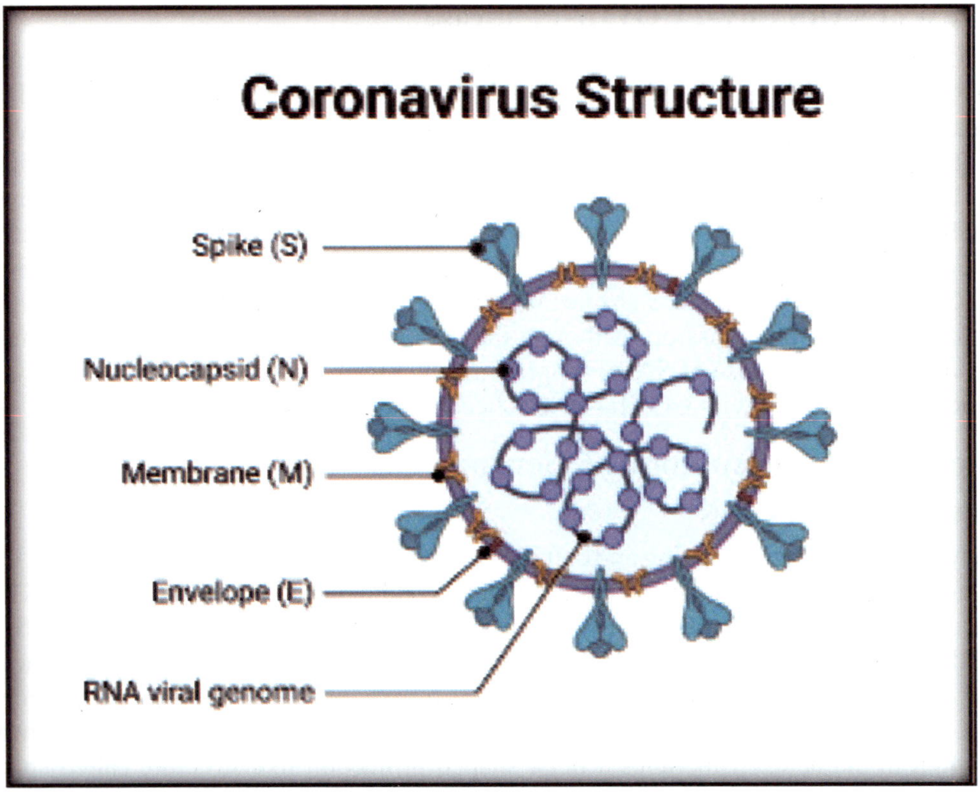

Figure 1- Coronavirus structure

(image taken from Biophysical society for understanding purpose only)

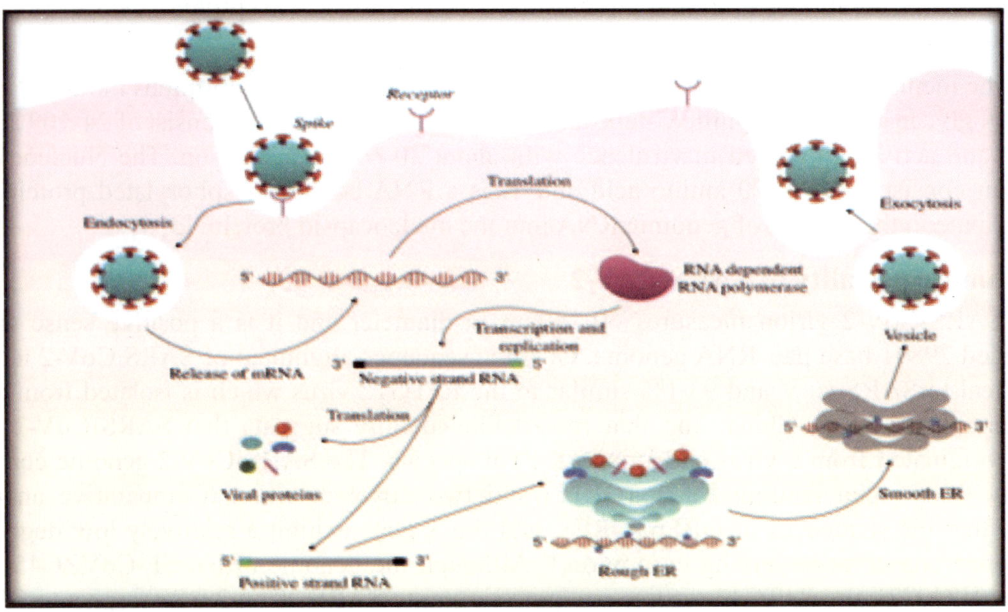

igure 2: - Genomic structure of Coronavirus

(image taken from geneticeducation.co.in for understanding purpose only)

Infectious quality of SARS-Cov-2 Virus

The binding of SARS-CoV-2 spike protein to its cell surface receptor angiotensin converting enzyme 2 (ACE2), initiate viral entry into type 2 pneumocytes in the human lung. The s protein play an important role in the initial transmission and infection of SARS-CoV-2. Spike protein consist of two domains S1 domain which is located at the N-terminus of the protein and mediates binding to ACE2 enzyme, S2 domain located at the C-terminus and it promotes fusion of the virus membrane with the cellular membrane of the host cell. The receptor binding domain is a subdomain of S1domin of spike protein made up of 424-494 amino acids that is direct contact with extracellular binding site on ACE2 which is known as Peptidase domain (PD). There are two cleavage site which is essential for the virus entry these are arginine R667 and R797. Numerous cellular proteases, including cathepsin L, trypsin, elastase, serine transmembrane proteases (TMPRSSs), and factor Xa, can cleave the S sequence at these two sites. Cleavage at both S protein sites is critical for S1 binding to ACE2 and essential for membrane fusion.

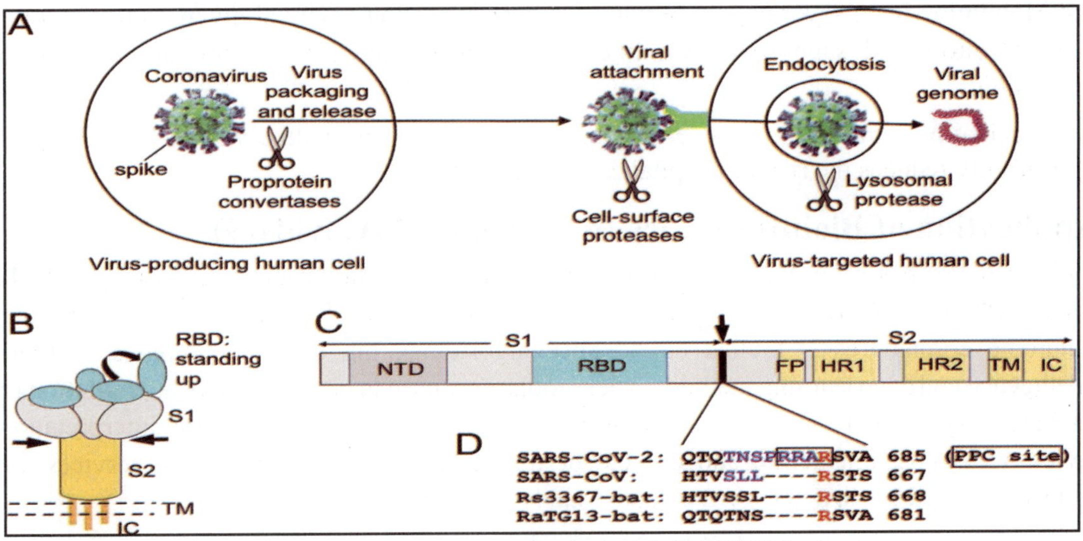

Figure 3: - Cell entry mechanism of SARS-CoV-2 virus

Symptoms, Diagnosis and Treatments

Covid 19 caused by SARS CoV-2 virus. Covid 19 symptoms, diagnosis and treatment are vary to person to person but some common include:

Symptoms of COVID 19

- **Respiratory symptoms:** cough, shortness of breath, difficulty in breathing
- **Systemic symptoms:** fever, chills, fatigue, muscle or body aches
- **Gastrointestinal symptoms:** nausea, vomiting, diarrhoea
- **Neurological symptoms:** loss of taste or smell (anosmia), headache
- **Severe symptoms:** chest pain, confusion, inability to awake or stay awake.
- Symptoms are mostly appeared in 2 to 14 days after the entry of virus in body. Only severe cases may require hospitalization.

Diagnosis of COVID 19

There are three methods of testing included

- PCR tests
- Rapid antigen tests
- Antibody tests

PCR test is most common and reliable test. All three tests are very sensitive, give quicker and reliable results.

Treatment of COVID 19

Treatment of patients are varied on the basis of their severity.

There is home care treatment for mild patients like take proper rest and keep body hydrated, eat healthy food. There are some antibiotics also provided for patients like Paxlovid, Remdesivir, and Molnupirovir. In sever condition where is hospitalization essential given corticosteroids, IL-6 inhibitors and supportive care like supplementary oxygen support and mechanical ventilation.

Vaccination remains the most effective way to prevent sever illness from COVID 19. It significantly reduces the risk of hospitalization and death. [3,4]

Application of Bioinformatics In Sars-Cov-2 (Covid-19)

Bioinformatics has been playing important role in the scientific progress to fight against the widespread of the Coronavirus disease in 2019, which is caused by severe acute respiratory syndrome coronavirus-2, fuelled by advance algorithms, big data technologies, artificial intelligence and deep learning methods. These tools were used in genomic analysis, evolutionary tracking, phylogenetic analysis, epidemiological analysis, protein structure interpretation, review virus host interaction and clinical performance. There are many web services and software applied for SARS CoV-2 database analysis. [1,2]

There are many methods used in COVID 19 understanding including

Genomic sequencing and analysis

Bioinformatics tools have been necessary for analysing the genomic sequences of SARS CoV-2. Bioinformatics tools have enabled rapid sequencing and analysis of the SARS-CoV-2 genome with the use of Next generation sequencing (NGS) and Third generation sequencing (TGS) technologies, which is important for understanding the virus's structure, function, and evolution. Tools like CoV-seq help in annoting genome sequence and identify genetic variants, while another database like GISAID provide access to a vast repository of viral genomes. [5,6,7,8] This methods allow researcher to find out:-

Obtain high quality genomic data, genome assembly and annotation

By swiftly producing complete genome sequences with the help of tool like CoV-Seq scientist can identify genetic variants and mutation that may affect the transmissibility or virulence of virus over the time and providing better understanding of the virus's evolutionary dynamics.

Next Generation Sequencing (NGS)

This technology used for fast sequencing of SARS CoV-2 genome. It used to track mutation and variant which is rise over the time. By analysing more than 9 million genetic sequences from multiple countries, scientist can identify variant specific characteristics that related with the infection.

Variant calling, tracking and mutation analysis

There are many tools that are used to detect and characterize mutation within the viral genome for providing better understanding of how specific variant may influence transmissibility and virulence of virus.

For variant calling, tracking and mutation analysis we used tools like-

Genome analysis toolkit: A versatile toolkit for variant calling and analysis use to identify mutation in SARS CoV-2 genomes.

SAMtools: A suite tools for manipulating and analysing sequence alignment (SAM/BAM) files essential for variant calling and analysis.

Pangolin: A tool for assigning lineage to SARS CoV-2 genome based on their genetic sequences.

Nextstrain: A powerful platform for real time tracking of pathogen evolution. It visualizes the spread of variants and their associated mutation through interactive phylogenetic trees.

COVID-19 Genomics UK (COG-UK): A UK based consortium that has been instrumental in sequencing and analysing SARS-CoV-2 genome. It provides insight into the evolution of the virus and the emergence of new variant.

Variant hunter: This is a user-friendly web application that allows for systematic monitoring of COVID 19 virus at both global and regional level. This software analyses amino acid changes over a specific time interval, geographical areas and provide lineage independent specific samples.it also help to track the prevalence of mutations and identify new viral variant from large dataset.

MicroGMT: Micro genomics mutant tracker is like a python package. It facilitates the detection of mutation from user provide sequence data.

CDC'S national SARS-CoV: 2 strain surveillance- The CDC utilize genomic inspection pipeline to pile up and analyse genetic sequence data from public health laboratories across the U.S and identify the circulating variants, inform public health responses to potential threats posed by new mutation.

These all tools are excel in variant discovery, analysis, excel in accuracy, speed and also work on low coverage sequencing data.[5,6,7,8]

Epidemiology Modelling

Bioinformatics has been a turning point in recognizing and battle against COVID-19, with incorporating genomic data with epidemiological information, so scientists can track the virus's mutations and see how they correspond with outburst. This helps in predicting future waves, identifying new variants, and guiding public health responses. For example, tools like Nextstrain gives real-time tracking of pathogen evolution, and they have been invaluable during the pandemic. By visualizing the virus's genetic changes and linking them to geographic data, researchers can identify hotspots and predict where the virus might spread next. This information supports targeted interventions, ensuring that resources are allocated efficiently and effectively.

Modelling evolutionary dynamics and Predictive modelling

With the help of computational models, researchers can simulate how mutations may affect viral behaviour and public health outcomes so they can guiding containment strategies. They utilize machine learning algorithms to predicted potential outbust based on genomic data and historical trends.

Understanding immune response

For understanding immune response bioinformatician analysing how mutations affect immune responses supports in predicting vaccine efficacy against new arising variants.

Susceptible- Exposed-Infection removed models

These models are mostly used to predict the pandemic curve by simulating different stages of infection spread. They incorporate data on transmission rates, recovery rates, and other epidemiological factors

Machine learning algorithm

Advanced machine learning techniques analyse genetic and epidemiological data to predict which variants are likely to become dominant in the coming weeks. For instance, a model developed from 9 million sequences has shown strong predictive power for variant spread. AI- driven models can analyse large datasets to predict how specific mutation affect viral properties so they can design vaccine and new therapeutics ideas. Machine learning algorithms modernized the processing of huge amount of genomic data and making it easier to identify viral evolution.

Phylogenetic Analysis

Bioinformatics tools validate researchers to build phylogenetic trees that demonstrate the evolutionary relationships between different SARS-CoV-2 strains, so that we can understand the relationships between different virus strains to trace transmission pathways. This include:-

Tracking lineage

Lineage tracking in covid 19 include observation of the evolution of SARS CoV-2 virus by analysing its genetic sequence so that scientist can trace the lineage of variants because as the virus replicates it accumulates mutation leading the outbreak of new virus and its transmissibility.

Identify natural host

Phylogenetic studies support that bats are probably the natural hosts for SARS-CoV-2, helping to investigate its genesis in humans.

Phylogenetic trees

Bioinformatics tools like PAUP and IQ-TREE are used to developed phylogenetic trees that demonstrate the evolutionary relationships between different variants. This trees helps in understanding how variants spread geographically and temporally.

- **Lineage assignment:-**Software such as Pangolin assigns lineages to genomic sequences based on established nomenclature, aiding in tracking specific variants as they emerge and spread.[5,6,7,8]

Understanding virus host interaction

Single cell RNA sequencing

Single cell RNA sequencing (scRNA-Seq) is a very capable technique that allow scientist to profile gene expression at the single cell level. This powerful technology has transformed understanding of cellular heterogeneity and has opened up new path of research in various field including biology, medicine and neuroscience. This technique provides insights into immune responses at a cellular level, helping identify how the virus affects different cell types in COVID-19 pandemic time. [9,10]

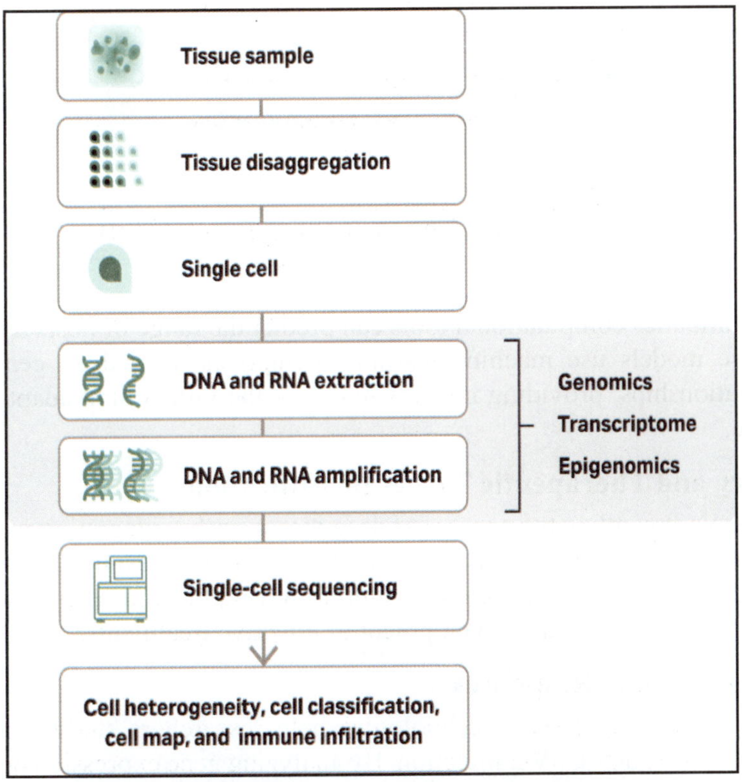

Figure 4: -Workflow of single cell RNA sequencing method

Analysis of Host and Viral RNA Sequencing Data

Bioinformatics methodology provide comprehensive analyses of RNA sequencing data from infected human cells, so they can identified differentially expressed genes and pathways specific to SARS-CoV-2 infection, such as those involved in immune responses and cellular processes. This helps in pointing out how the virus manipulates host cellular machinery to establish infection and evade immune detection.

Prediction of Protein-Protein Interactions (PPIs)

Bioinformatics tools are important for predicting interactions between viral proteins and host proteins, using algorithms that analyze structural similarities and sequence motifs, researchers can recognize potential binding partners between SARS-CoV-2 and human proteins. Such insights are important for understanding the viral pathogenesis and could lead to the identification of new drug targets.

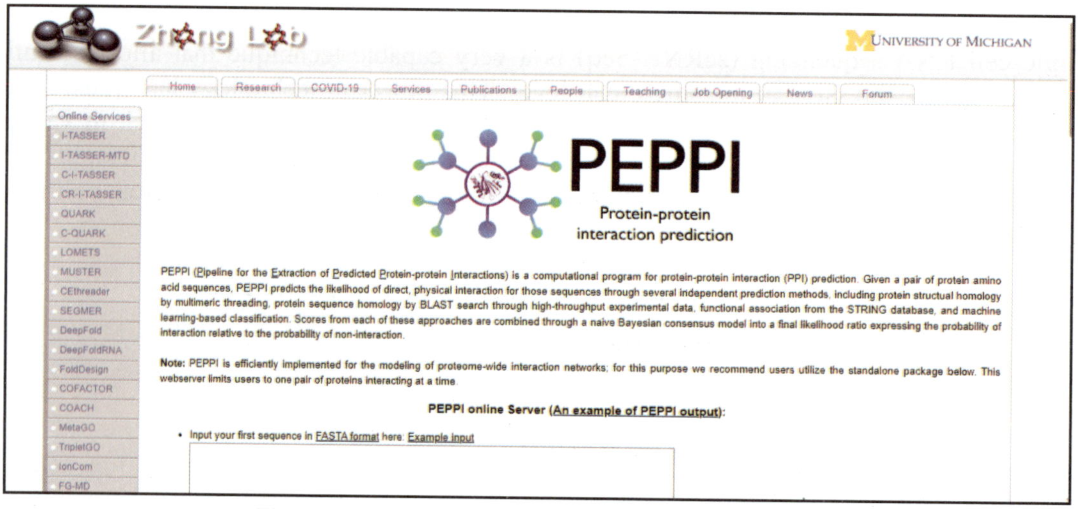

Figure 5:- PROTEN PROTEIN PREDICTION TOOL

Host Range Prediction

Advanced bioinformatics computational tools can predict the range of hosts that SARS-CoV-2 can infect. These models use machine learning techniques to analyze genomic traits and phylogenetic relationships, providing insights into how the virus might adapt to new hosts or evolve over time.

Drug Discovery and Therapeutic Target Identification

Bioinformatics tools streamline the process of drug discovery by integrating various datasets to identify potential therapeutic targets. Computational-aided drug design (CADD) methods can predict which compounds might effectively inhibit viral replication or enhance host immune responses, thereby accelerating the development of effective treatments.

Understanding Immune Responses

Next-generation sequencing (NGS) technologies help researchers study how host genetics influence responses to SARS-CoV-2 infection. By analyzing gene expression profiles, scientists can determine how variations in host genes affect susceptibility to infection and disease severity, which is vital for developing personalized medicine approaches. [9,10]

Practical time

How to use protein protein interaction tools

Step 1: First open protein protein interaction prediction tools page on web browser.

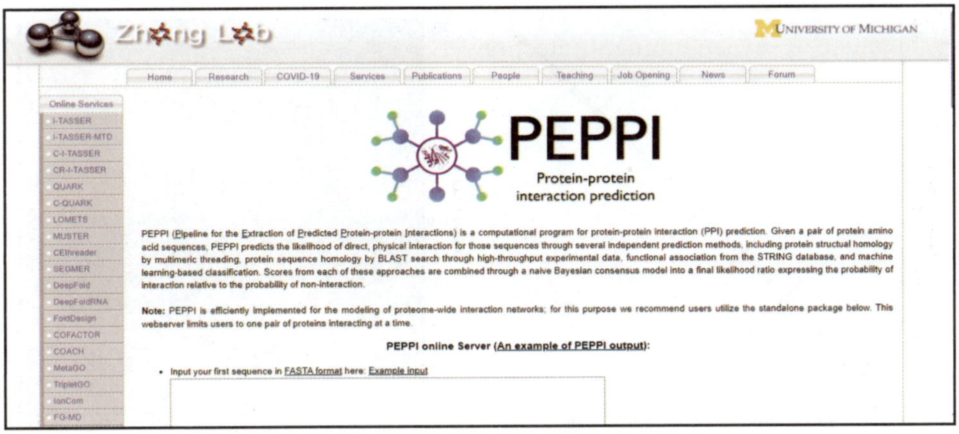

Step 2: After open this page choose deep fold option from online services bar

Step 3: After choosing deep fold option you can see this type of page on your screen

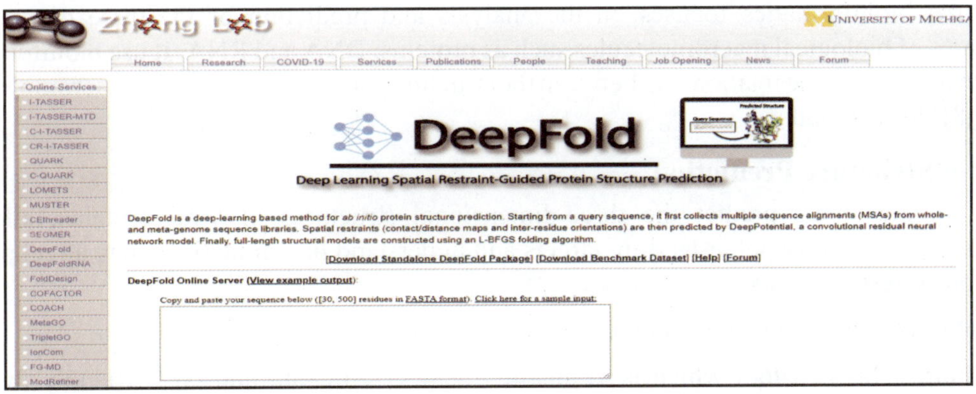

Step 4: After open this page you fill all the information which they asked

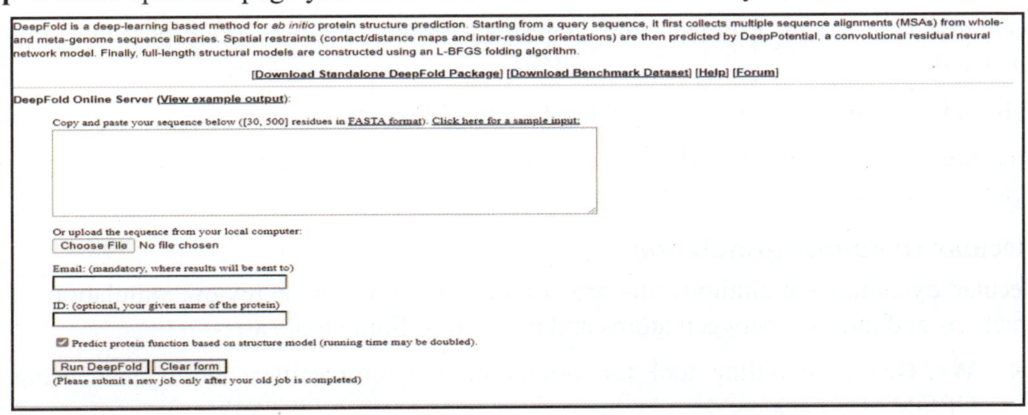

Step 5: Than run the software after sometime you will received result of your sample in your mail id and analyse the result

Step 6: You can find this type of result

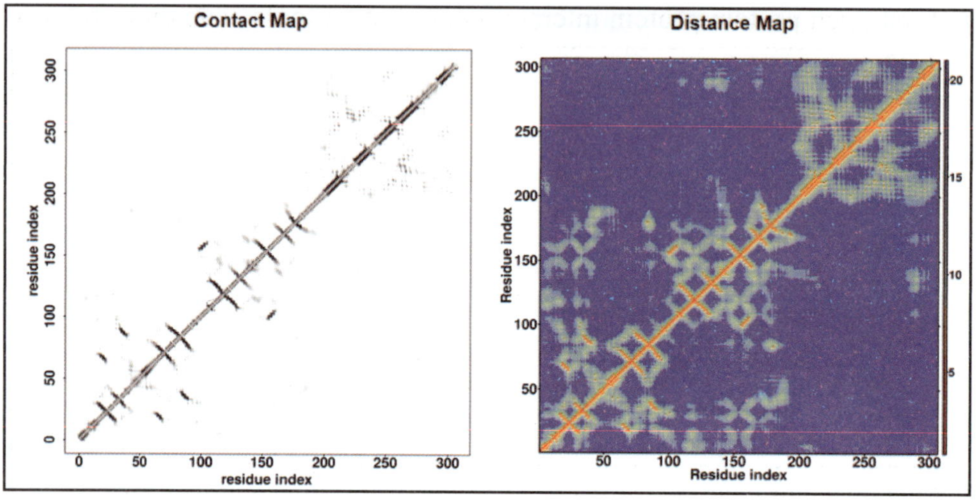

Structural bioinformatics

Structural bioinformatics focuses on the analysis and prediction of the three dimensional structures of biological macromolecules such as proteins, DNA and RNA. It use bioinformatics tools to understand the relationship between the structure and functions of these macromolecules. [5,6,7,8]These includes:-

Protein Structure Prediction

Bioinformatics tools predict the three-dimensional structures of viral proteins, allowing researchers to identify potential drug targets and understand how mutations may affect protein function. **These tools are**

AlphaFold2- This is an Open-source code for AlphaFold

OpenFold- This is a tool which is Trainable, memory-efficient, and GPU-friendly PyTorch reproduction of AlphaFold 2.

HelixFold- This is an Efficient and Improved Implementation of AlphaFold 2 using PaddlePaddle.

ColabFold- This is web server of AlphaFold2 using MMseqs2.

AlphaFold protein structure- AlphaFold is an AI system developed by DeepMind that predicts a protein's 3D structure from its amino acid sequence

Molecular Dynamics Simulation

Molecular dynamics simulation tools are software programs that review simulation of the interactions and motions between atoms and molecules. Some tools are:-

- **WebGro :** an online tool for simulating macromolecules and ligands using the GROMACS package.
- **LAMMPS :** a code for materials modelling with potentials for solid-state and soft matter systems

These simulations help predict how viral proteins interact with potential vaccine candidates, providing insights into stability and efficacy.

Drug Discovery and Vaccine Development

Drug discovery and vaccine development is a complex and tedious process that involves identifying, developing and testing new medicines to treat disease. [11,12] It is typically following these steps:

- Target identification and validation
- Hit identification
- Lead optimization
- Preclinical development
- Clinical trials
- Regulatory approval and post market surveillance

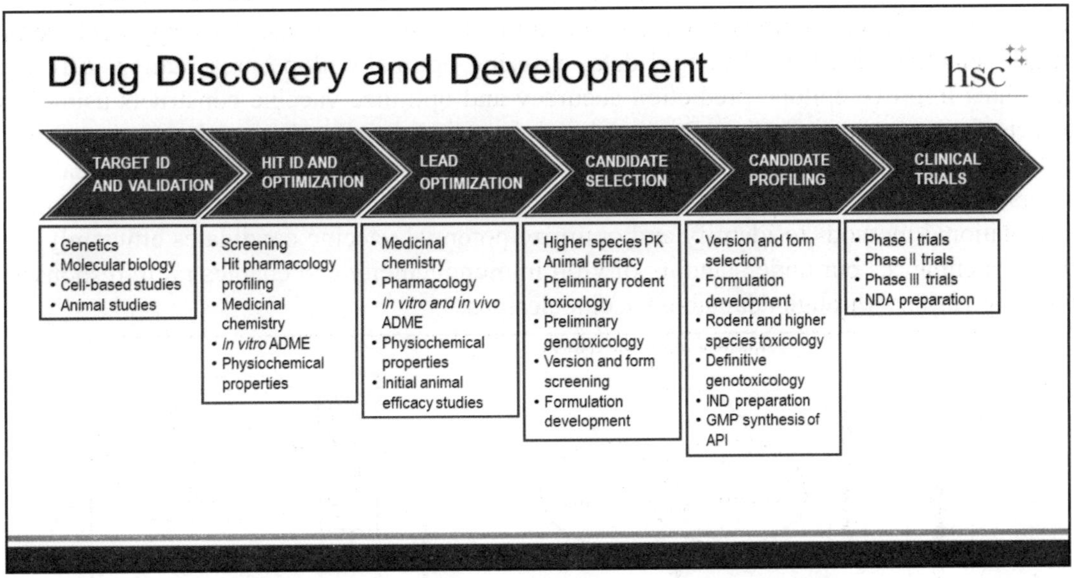

Figure 7 :-Drug discovery and vaccine development steps

Reverse Vaccinology

Reverse vaccinology has become a crucial strategy in developing COVID-19 vaccines. This method uses computational techniques to identify potential vaccine candidates from the viral genome. It enables the swift design and assessment of vaccine constructs, especially in response to emerging variants such as Omicron. This approach uses genomic data to identify antigens that could be targeted by vaccines, leading to the development of effective immunizations against COVID-19. Reverse vaccinology starts with identifying immunogenic epitopes from the SARS-CoV-2 genome. Bioinformatics tools predict epitopes that can bind to Major Histocompatibility Complex (MHC) molecules, crucial for T-cell recognition. Studies have identified multiple MHC I and MHC II epitopes from the spike protein of SARS-CoV-2, essential for eliciting immune responses. Besides T-cell epitopes, B-cell epitopes are identified to stimulate antibody production. Tools like ABCpred and ElliPro predict linear and conformational B-cell epitopes for inclusion in vaccine construct. After identifying potential epitopes, they are assembled into a vaccine construct. Selected epitopes are linked together, often with adjuvants to enhance immunogenicity. For example, a recent study constructed a multiepitope vaccine candidate by combining various T-cell and B-cell epitopes with an adjuvant protein. The vaccine construct undergoes various in silico analyses to assess its stability, solubility, and potential immunogenicity before experimental validation. Understanding the vaccine candidate's structure is crucial for predicting its interaction with immune receptors. Tools like AlphaFold and RaptorX predict the three-dimensional structures of the vaccine constructs, helping understand how they will present epitopes to the immune system. Docking studies evaluate how well the vaccine candidate binds to immune receptors such as Toll-Like Receptors (TLRs) or ACE2, providing insights into its potential efficacy in eliciting an immune response. Reverse vaccinology enables the assessment of vaccine candidates against different SARS-CoV-2 variants. By comparing the mutation profiles of epitopes between variants (e.g., Omicron BA.1 and BA.2), researchers can estimate how effectively a vaccine candidate might work against these strains. This adaptability is crucial given the rapid evolution of SARS-CoV-2. Reverse vaccinology is often combined with other computational methods. This approach refines antigen selection based on structural data, ensuring optimal epitope presentation. Advanced algorithms improve epitope prediction accuracy and optimize vaccine constructs using large datasets from previous studies. Predicted vaccine candidates undergo laboratory testing to confirm their ability to induce immune responses in vitro before progressing to clinical trials. Reverse vaccinology has significantly accelerated COVID-19 vaccine development by utilizing computational methods to identify and optimize potential vaccine candidates efficiently. This approach enhances our understanding of viral immunogenicity and enables rapid responses to emerging variants through adaptable vaccine designs.

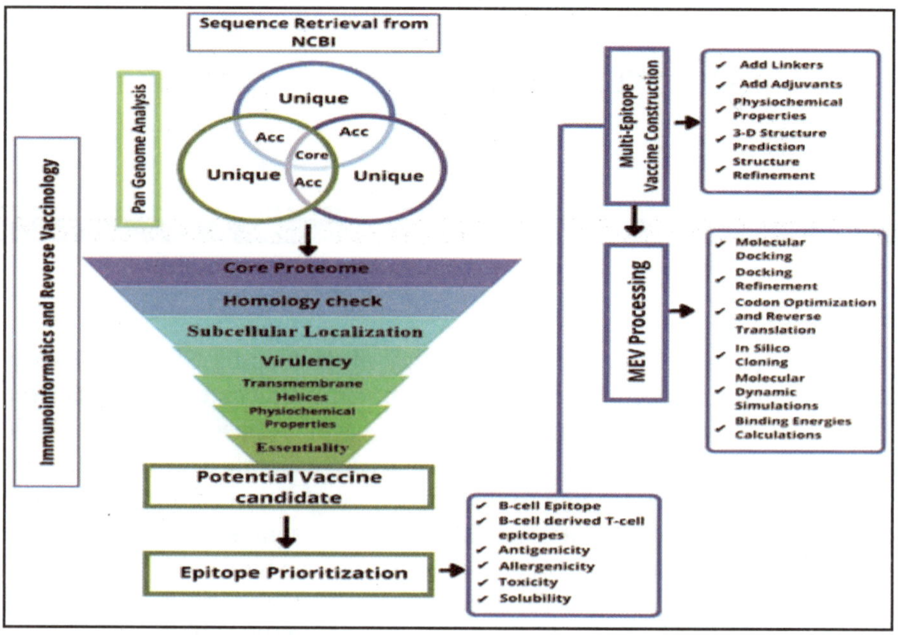

Figure 9:-Immunoinformatics and reverse vaccinology

Epitope Prediction and Selection

Bioinformatics facilitates the identification of potential epitopes—specific parts of antigens that trigger an immune response. Advanced computational methods, such as deep learning architectures combining Convolutional Neural Networks (CNN) and Long Short-Term Memory networks (LSTM), are employed to predict B-cell and T-cell epitopes from the SARS-CoV-2 proteome. These methods improve prediction accuracy by analyzing sequence data and structural information, enabling researchers to select the most promising candidates for vaccine development. [11,12]

Multi Epitope Vaccine Design

The integration of immunoinformatic techniques allows for the design of multi-epitope vaccines that can induce robust immune responses. By predicting multiple epitopes that can stimulate both B-cell and T-cell immunity, researchers can create vaccines that are more likely to provide comprehensive protection against the virus. This approach has been shown to yield candidates with high antigenicity and low toxicity, which are critical for safe and effective vaccines.

Toxicity and Allergenicity Assessment

Bioinformatics tools such as ToxinPred and AllerTop are utilized to evaluate the safety profiles of vaccine candidates by predicting their toxicity and potential allergenic responses. This step is essential in ensuring that the selected epitopes do not provoke harmful reactions in vaccinated individuals, thereby increasing the likelihood of successful vaccine deployment.

Codon Optimization and Cloning

Once potential vaccine candidates are identified, bioinformatics aids in optimizing codons for efficient expression in host systems. Tools like SnapGene facilitate cloning processes, ensuring that the vaccine constructs are suitable for production in laboratory settings. This optimization is vital for achieving high yields of the desired protein or peptide.

Simulation of Immune Responses

Bioinformatics also enables the simulation of immune responses using platforms like IMMSIM. These simulations help predict how well the immune system will respond to the vaccine, allowing researchers to make informed adjustments before moving into clinical trials.

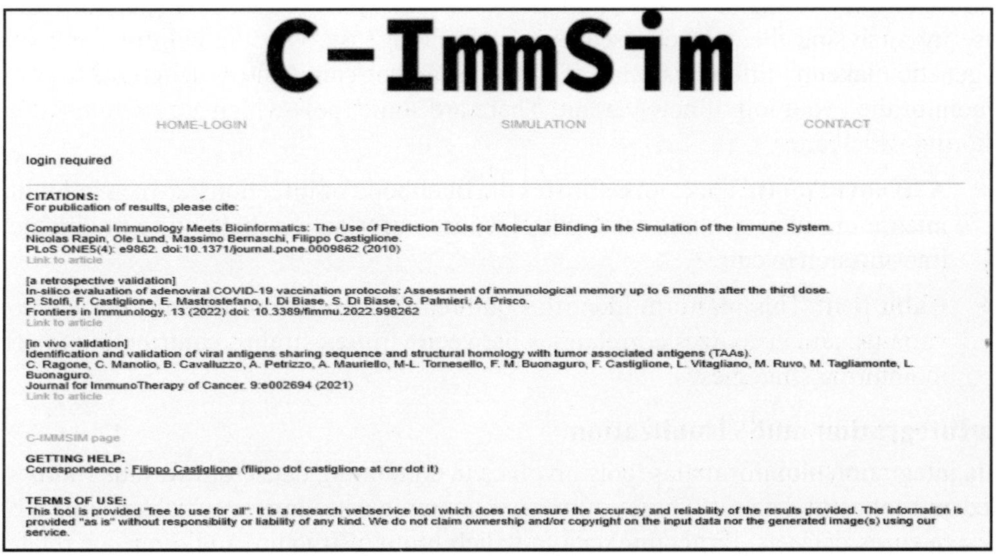

Figure 10: -Immsim immunology server

Practical Time

How to use tools for B-cell/ T-cell Epitope prediction tools

Step 1: - Take your sample file with Protein data bank

Step 2: - Get sequence or FASTA file from MMDB

Step 3: -Open IEDB analyse resources tool web page for epitope prediction

Step 4: - Fill all the requirements

Step 5: - Submit your sample and analyse your result

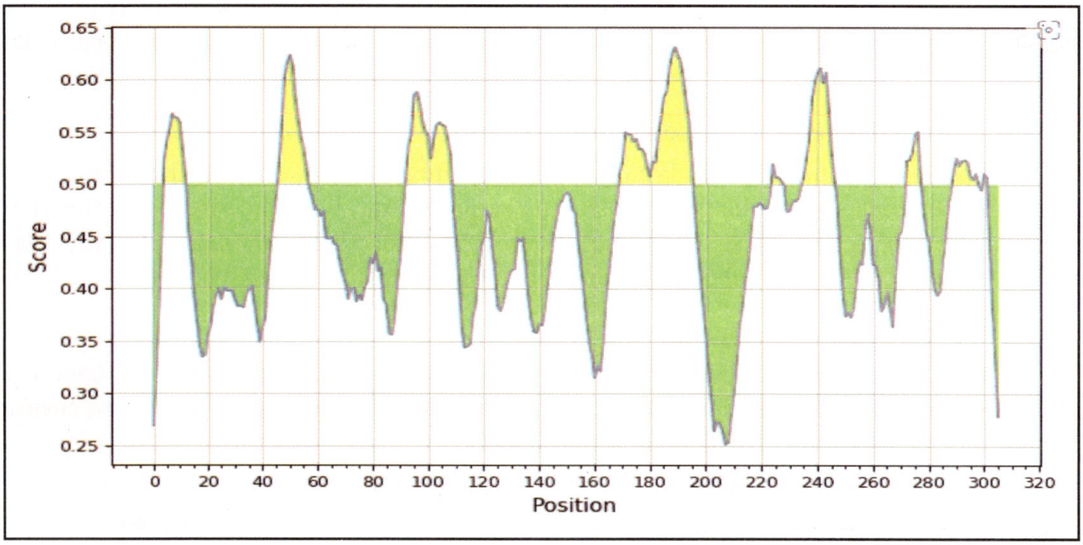

This type of results you can find from epitope prediction tools

Transmission Monitoring

Transmission monitoring is a bioinformatics tool that used to track the speed of infectious disease by analysing their genetic sequences of pathogens. With the help of comparison of their genetic makeup, different strains of virus researcher can identify pattern of transmission and monitor the evolution of new variant. There are some tools which are used in COVID-19 monitoring which are:-

- **A2B covid tools:** This tool estimates the likelihood of infection between individuals by integrating viral genomic data with location information, helping to predict potential transmission events
- **Calm belt:** This platform identifies outbreak transmission patterns, detects potential variants, and visualizes correlations between virulent strains, contributing to effective monitoring strategies.

Data Integration and Visualization

In data integration bioinformatics tools involves in combining data from various sources into a unified and coherent data set. These sources are genomic data sources which provide DNA and RNA sequences datasets, Proteomic data in which protein structure and sequence related data sets are present, Metabolic data there are biochemical and reactions data sets are present, clinical data gives information about patients records and medical histories. In data visualization this is a process in which data representing in the form of graphs to make it easier to understand and interpret. These both methods have some applications like: -

- **Rapid data availability-** Researchers can access a wealth of genomic sequences from around the world, facilitating swift analysis and response to new variants.
- **Cross institutional research-** Collaborative platforms enable scientists to pool resources and expertise, enhancing the overall understanding of SARS-CoV-2 evolution
- **Big data analysis-** Platforms like nCov2019 aggregate data on SARS-CoV-2 strains, allowing for real-time visualization of infection spread across different regions. This integration aids public health officials in making informed decisions about interventions
- **AI based prediction system -** Tools such as Blue Dot use artificial intelligence to detect outbreaks and visualize the spread of the virus in real-time, enhancing situational awareness during the pandemic

Network Pharmacology

Network pharmacology is a new methodology that uses computational methods to systematically catalogue the molecular interaction of a drug molecule in a living cell. Its aim to understand drug action and their interaction with the multiple targets and to understand disease working pathways. Network pharmacology has appeared as a powerful approach in the fight against COVID-19, using computational methods to understand the complex interactions between drugs, biological targets, and disease pathways. [13,14,15]

Applications of Network Pharmacology in COVID-19

Network pharmacology supports discovering of potential therapeutic compounds from traditional medicines and natural products. In **Compound-Target Interaction** researcher analysing databases like TCMSP and Gene Cards, researchers can point active compounds that interact with specific targets related to COVID-19. For example, some studies have identified compounds such as naringenin and quercetin from various plants that show potential against SARS-CoV-2 by targeting important proteins which is involved in viral replication and immune response. COVID-19 is a complex disease, and network pharmacology helps in understanding how multiple targets can be modulated simultaneously with the help of Pathways Analysis methods. Researchers use tools like KEGG (Kyoto encyclopaedia of Genes and Genomes) and GO (Gene Ontology) to enhance quality of analysis to review how identified compounds affect different biological pathways which are related to inflammation, immune response, and viral infection. For example, studies have shown that certain herbal formulations can regulate pathways like IL-17, PI3K-Akt, and JAK-STAT, which are crucial in managing cytokine storms seen in severe COVID-19 cases. Network pharmacology findings are often validated through experimental studies with the help of the methods of In Vitro Confirmation. After investigating active components and their mechanisms, researchers performed laboratory experiments like ELISA, PCR to confirm the biological effects of these compounds on COVID-19-related pathways. This type of validation support and strengthens the evidence for using specific herbal medicines or compounds in clinical settings. Network pharmacology offers a comprehensive approach to exploring potential treatments for COVID-19 by identifying active compounds, understanding multi-target mechanisms, predicting drug interactions, and validating findings through experimental methods. This integrative strategy enhances our understanding of COVID-19 therapeutics and supports the development of effective treatment options.

Advantages of Drug Repurposing

Drug repurposing is the methods of using an already discovered drug for a new treatment that was not originally intended use. This can be very cost-effective way to find new treatment plan for disease, as it avoids the expensive and time-consuming process of developing a new drug from scratch.it is important tool in the fight against disease and play an even greater role in the future. There are many advantages of drug repurposing. [13,14,15] This includes: -

Time and Cost Efficiency

Drug repurposing permits researchers to recognize new uses for existing medications, notably lowering the time and costs associated with developing new drugs from scratch. This approach can avoid several phases of clinical trials, as existing drugs have already undergone safety assessments.

Rapid Data Utilization

The COVID-19 pandemic conduct to an unparalleled accumulation of biological data regarding SARS-CoV-2, including genomic, transcriptomic, and proteomic profiles. Bioinformatics tools have been used to extract this data effectively, promote the identification of potential therapeutic targets and candidate repurposable drugs.

Integration of AI and Machine Learning

Advanced computational tools, including artificial intelligence (AI) and machine learning (ML), have been working to analyze huge datasets and predict which existing drugs could be effective against COVID-19. To illustrate, machine learning algorithms have successfully identified promising candidates like Baricitinib and Dexamethasone through analysis of biological pathways and drug interactions.

AI-repurposed drugs for COVID-19

There are many challenges arise from the complexities of drug repurposing [13,14,15] which includes: -

Evolving Nature of SARS-CoV-2

The rapid mutations of the SARS-CoV-2 virus gives complications in the development of effective treatments. For these continuous variations or mutations in variants can alter transmission rates, disease severity, and immune escape capabilities, which involve complicated adjustments in treatment strategies and regulatory approaches.

Regulatory Framework Limitations

Right now, there is an inadequacy of an absolute regulatory frame work design to intercept the specific challenges with the drug repurposing. Traditional drug approval methods might not sufficiently help the complexities of repurposing exiting drug for new uses those complexities resulting in delays and equivocation in the approval procedure.

Clinical Trial Requirements

Repurposed drugs frequently used new clinical trials to confirm their safety and effectiveness in different patients' groups, groups which were initially approved. This procedure involves arranging phase trials to evaluate new formulations and combinations of drugs. This process can be both resource intensive and time consuming. Additionally, if a repurposed drug is combined with another agent, regulatory framework requires additional trials to access their potential and toxicities.

Data Quality and Availability

AI-manage drug repurposing remarkably based on high-quality data. Although, obstacles like data harmonization and limited access to comprehensive clinical datasets obstruct the process of making cultured decisions about repurposed drugs. The absence of broadly available patient data made more complicates these issues, especially considering the variability in patient responses in real time world scenario.

Intellectual Property Issues

Pharmaceutical companies might encounter inhibition to develop identified repurposed drugs because of current intellectual property protections. This situation can turn down the company's eagerness to invest in additional clinical development if they believe they cannot recover their costs or achieve market exclusivity.

Complexity of Biological Mechanisms

Grasping the molecular mechanisms through which repurposed drug's function is essential but often complex. The indirect nature of target identification can create uncertainty about the suitability of a drug for a new indication, further complicating regulatory evaluations.

Addressing these challenges demands collaboration among stakeholders, including regulatory bodies, pharmaceutical companies, and researchers. Creating standardized protocols for assessing repurposed drugs could improve their incorporation into clinical practice and speed up access to effective treatments for COVID-19. As AI continues to evolve, its potential to streamline drug discovery must be paired with robust regulatory frameworks capable of adapting to these innovative methods.

Practical time: - some important tools for practice

String (search tool for protein-protein interaction analysis) . String incorporates automated text mining and computaional predictions, making it a benificial resource for network enrichment analysis

Step 1: -Take your protein sample

Step 2: -Open string database tool

Step 3: - Submit your sample

Step 4: - Analyse your result

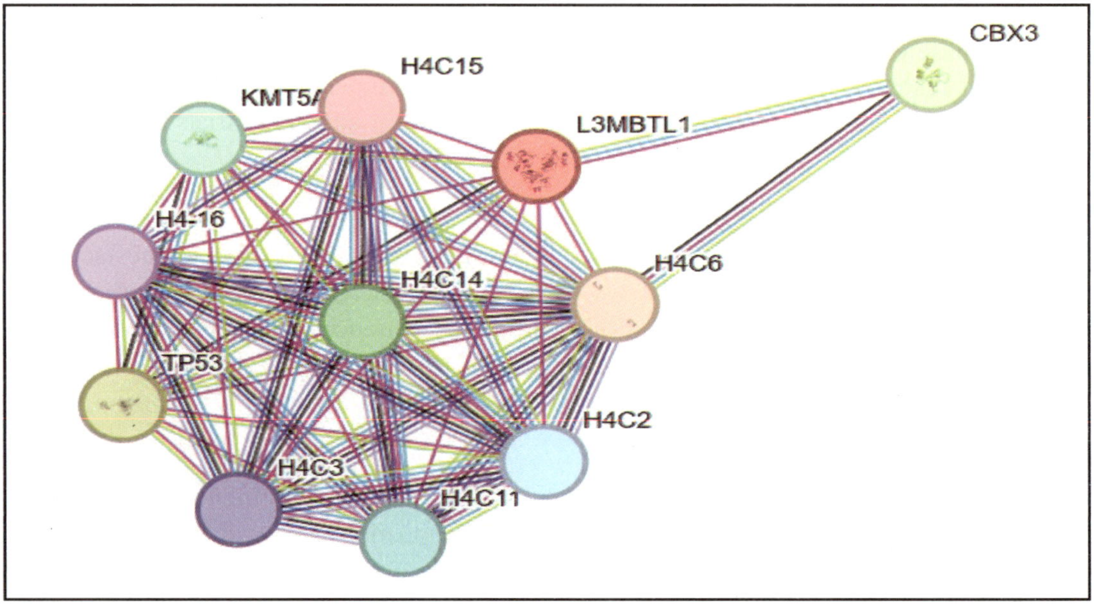

This type of result you can found analyse your result

Scope in Bioinformatics

There are many scopes in the field in bioinformatics. Current era of the worldis now appropriate for bioinformatics. Scope of bioinformatics includes:

- **Healthcare industries:** bioinformatics scientist contribute to their work in healthcare industries in the form of making medical diagnostics handy with accuracy, innovate personalized medicine for better treatment with precision and clinical decision support system in healthcare setups.
- **Pharmaceutical companies, drug discovery and designing:** Bioinformatics plays an important role in pharm companies with the discovery of drug designing, target identification, virtual screening libraries and optimization of therapeutic compounds in pharmaceutical research and development.
- **Research and development:** Bioinformaticians discover multi-omics techniques including genomics, proteomics, metabolomics, epigenetics, system biology and structural biology and give speed to research sector and developed great knowledge.
- **Data science:** Bioinformatics supports data science careers with the application in analyzing large biological datasets, developing predictive models, and extracting actionable insights.
- **Clinical research:** Bioinformatics help in clinical research and empowering this field. In clinical research they do biomarker discovery, disease modelling, pharmacogenomics studies, network pharmacology, and contribute to improved patient's outcomes and give precise healthcare treatment.
- **Software development and computational biology:** Bioinformatics experts developed software for computational biology tools apply mathematical and computational techniques for analyzing and study biological datasets, enhance research capabilities,

uncovering biological pathways, principles and mechanisms, facilitates data driven decision making ability.

Limitation of Bioinformatics

When bioinformatics gives us new scopes in many filed so there are some limitations also coming with this. Some limitations includes: -

- **Data quality:** Research are depends on data generated by various research. This data should be good in quality for better results but sometime various experimental techniques gives data which is vary in quality and accuracy, this type of data gives false results those are not good for research.
- **Computational complexity:** In the research and for good results analyzing large datasets require significant computational resources which leading to challenges in the form of processing time and storage.
- **Algorithmic limitations:** In bioinformatics there are many algorithms present. These algorithms may not always be able to giving accurate and predict complex biological datasets or can't interpret highly heterogeneous data.
- **Biological data understanding:** Bioinformatics tools integrates data form many sources like multi omics techniques and provide correlation and patterns but it may not always accurate and can be challenging due to difference in data formats and experimental biological mechanisms and protocols.
- **Ethical and legal issues:** There are many ethical issues which are related to concern of privacy, data security and ownership of biological data that pose ethical and legal challenges in research and applications.

Bioinformatics combine biology with computational methods, enhance researcher strength to analyse and interpret biological datasets with the accuracy and speedily. It facilitates genome sequencing, protein structure prediction and epidemiology analysis, fasting scientific discovery and boost innovation. Other than this innovative impact sometime bioinformatics faces challenges such as data quality issues and computational complexity. Through interdisciplinary collaboration, bioinformatics continue to drive advancement in fields like medicine, agriculture and environmental science.

Refrence

1. Can T. Introduction to bioinformatics. Methods Mol Biol. 2014; 1107:51-71. doi: 10.1007/978-1-62703-748-8_4. PMID: 24272431.
2. Bayat A. Science, medicine, and the future: Bioinformatics. BMJ. 2002 Apr 27;324(7344):1018-22. doi: 10.1136/bmj.324.7344.1018. PMID: 11976246; PMCID: PMC1122955.
3. Acuti Martellucci C, Flacco ME, Cappadona R, Bravi F, Mantovani L, Manzoli L. SARS-CoV-2 pandemic: An overview. Adv Biol Regul. 2020 Aug; 77:100736. doi: 10.1016/j.jbior.2020.100736. Epub 2020 Jun 17. PMID: 32773099; PMCID: PMC7832554.
4. Shi Y, Wang G, Cai XP, Deng JW, Zheng L, Zhu HH, Zheng M, Yang B, Chen Z. An overview of COVID-19. J Zhejiang Univ Sci B. 2020 May;21(5):343-360. doi: 10.1631/jzus.B2000083. Epub 2020 May 8. PMID: 32425000; PMCID: PMC7205601.

5. Tan M, Xia J, Luo H, Meng G, Zhu Z. Applying the digital data and the bioinformatics tools in SARS-CoV-2 research. Comput Struct Biotechnol J. 2023 Oct 1;21:4697-4705. doi: 10.1016/j.csbj.2023.09.044. PMID: 37841328; PMCID: PMC10568291.
6. Ma L, Li H, Lan J, Hao X, Liu H, Wang X, Huang Y. Comprehensive analyses of bioinformatics applications in the fight against COVID-19 pandemic. Comput Biol Chem. 2021 Dec;95:107599. doi: 10.1016/j.compbiolchem.2021.107599. Epub 2021 Nov 2. PMID: 34773807; PMCID: PMC8560182.
7. Kumar A, Asghar A, Dwivedi P, Kumar G, Narayan R, Jha R, Parashar R, Sahni C, Pandey S A Bioinformatics Tool for Predicting Future COVID-19 Waves Based on a Retrospective Analysis of the Second Wave in India: Model Development Study,JMIR Bioinform Biotech 2022;3(1):e36860,URL: https://bioinform.jmir.org/2022/1/e36860 DOI: 10.2196/36860
8. Ferrarini, M.G., Lal, A., Rebollo, R. *et al.* **Genome-wide bioinformatic analyses predict key host and viral factors in SARS-CoV٢- pathogenesis.** *Commun Biol* ٢٠٢١) ٥٩٠ ,٤). https://doi.org/١٠,١٠٣٨/s٠-٠٢٠٩٥-٠٢١-٤٢٠٠٣
9. Iuchi H, Kawasaki J, Kubo K, Fukunaga T, Hokao K, Yokoyama G, Ichinose A, Suga K, Hamada M. Bioinformatics approaches for unveiling virus-host interactions. Comput Struct Biotechnol J. 2023;21:1774-1784. doi: 10.1016/j.csbj.2023.02.044. Epub 2023 Feb 27. PMID: 36874163; PMCID: PMC9969756.
10. Khan MS, Yousafi Q, Bibi S, Azhar M, Ihsan A. Bioinformatics-Based Approaches to Study Virus-Host Interactions During SARS-CoV-2 Infection. Methods Mol Biol. 2022;2452:197-212. doi: 10.1007/978-1-0716-2111-0_13. PMID: 35554909.
11. Sunita, Sajid A, Singh Y, Shukla P. Computational tools for modern vaccine development. Hum Vaccin Immunother. 2020 Mar 3;16(3):723-735. doi: 10.1080/21645515.2019.1670035. Epub 2019 Dec 18. PMID: 31545127; PMCID: PMC7227725.
12. Ameen, Z. S., Mostafa, H., Ozsahin, D. U., & Mubarak, A. S. (2023). Accelerating SARS-CoV-2 Vaccine Development: Leveraging Novel Hybrid Deep Learning Models and Bioinformatics Analysis for Epitope Selection and Classification. *Processes*, *11*(6), 1829. https://doi.org/10.3390/pr11061829
13. Liu Z, Chen X, Carter W, Moruf A, Komatsu TE, Pahwa S, Chan-Tack K, Snyder K, Petrick N, Cha K, Lal-Nag M, Hatim Q, Thakkar S, Lin Y, Huang R, Wang D, Patterson TA, Tong W. AI-powered drug repurposing for developing COVID-19 treatments. Reference Module in Biomedical Sciences. 2022:B978-0-12-824010-6.00005-8. doi: 10.1016/B978-0-12-824010-6.00005-8. Epub 2022 Feb 23. PMCID: PMC8865759.
14. Blanco-González, A.; Cabezón, A.; Seco-González, A.; Conde-Torres, D.; Antelo-Riveiro, P.; Piñeiro, Á.; Garcia-Fandino, R. The Role of AI in Drug Discovery: Challenges, Opportunities, and Strategies. *Pharmaceuticals 16* ,٢٠٢٣, 891. https://doi.org/10.3390/ph16060891
15. Masoudi-Sobhanzadeh Y. Computational-based drug repurposing methods in COVID-19. Bioimpacts. 2020;10(3):205-206. doi: 10.34172/bi.2020.25. Epub 2020 Jun 18. PMID: 32793443; PMCID: PMC7416005.
16. Sunita, Sajid A, Singh Y, Shukla P. Computational tools for modern vaccine development. Hum Vaccin Immunother. 2020 Mar 3;16(3):723-735. doi: 10.1080/21645515.2019.1670035. Epub 2019 Dec 18. PMID: 31545127; PMCID: PMC7227725.

17. Ameen, Z. S., Mostafa, H., Ozsahin, D. U., & Mubarak, A. S. (2023). Accelerating SARS-CoV-2 Vaccine Development: Leveraging Novel Hybrid Deep Learning Models and Bioinformatics Analysis for Epitope Selection and Classification. *Processes*, *11*(6), 1829. https://doi.org/10.3390/pr11061829

18. J. Shang, Y. Wan, C. Luo, G. Ye, Q. Geng, A. Auerbach, & F. Li, Cell entry mechanisms of SARS-CoV-2, Proc. Natl. Acad. Sci. U.S.A. 117 (21) 11727-11734, https://doi.org/10.1073/pnas.2003138117 (2020).

19. GeeksforGeeks | A computer science portal for geeks

20. www.biophysics.org/education-careers/education-resources/covid-19-science-stories-and-resources/science-articles/coronavirus-structure-vaccine-and-therapy-development

21. geneticeducation.co.in/novel-coronavirus-covid-19-structure-genome-and-testing

15 Career And Scope of Bioinformatics

Introduction

The increasing volume of biological data generated by high-throughput technologies has necessitated the use of bioinformatics to extract meaningful insights. This field plays a crucial role in advancing scientific discovery and innovation across various domains such as healthcare, agriculture, and environmental science.

Bioinformatics is an interdisciplinary field that combines biology, computer science, mathematics, and information technology to analyze and interpret biological data. The modern biological sciences rely heavily on computational methods to process vast volumes of genomic, proteomic, and metabolomic information.

In recent decades, the rapid development of next-generation sequencing (NGS), computational modeling, and artificial intelligence (AI) has expanded the reach of bioinformatics beyond research laboratories to industries such as pharmaceuticals, agriculture, environmental science, and healthcare.This chapter explores the career opportunities, skill requirements, and future scope of bioinformatics, emphasizing its growing importance in data-driven biological research and sustainable development.

The growing reliance on computational approaches for solving biological problems has created a high demand for skilled bioinformaticians in academia, industry, and government sectors.Bioinformatics has emerged as a pivotal interdisciplinary field, bridging biology, computer science, and statistics to analyze and interpret complex biological data. This chapter explores the diverse career opportunities and scope within bioinformatics, emphasizing its applications in genomics, drug discovery, agriculture, and personalized medicine. We also discuss skillsets, emerging trends, and challenges faced by professionals in this field.

Evolution of Bioinformatics

The field of bioinformatics originated in the 1960s with the storage and analysis of protein sequences. The establishment of databases such as GenBank (1982), SWISS-PROT (1986), and the Protein Data Bank (PDB) laid the foundation for molecular data management.The completion of the Human Genome Project (2003) marked a turning point, creating a demand for professionals capable of managing, analyzing, and interpreting biological information. Today, bioinformatics has evolved into a core discipline that supports drug discovery, personalized medicine, and synthetic biology (Lesk, 2019).

Major Areas in Bioinformatics

Genomics and Transcriptomics

Bioinformaticians analyze DNA and RNA sequencing data to identify genes, mutations, and expression patterns. Applications include disease gene identification, biomarker discovery, and evolutionary studies.

Proteomics and Structural Bioinformatics

This area focuses on protein sequence analysis, 3D modeling, and interaction studies. Tools such as **BLAST**, **Clustal Omega**, and **AlphaFold** are widely used for protein structure prediction and comparative modeling.

Cheminformatics and Drug Discovery

Bioinformatics plays a major role in **virtual screening**, **molecular docking**, and **ADMET (Absorption, Distribution, Metabolism, Excretion, and Toxicity)** predictions for drug design. Integration with AI and machine learning enhances the efficiency of lead compound identification.

Systems Biology and Network Analysis

This involves studying interactions between genes, proteins, and metabolites to understand biological pathways and regulatory mechanisms. Systems biology contributes to the development of new therapeutic strategies and biotechnological applications.

Agricultural and Environmental Bioinformatics

Bioinformatics supports crop improvement, microbial genome analysis, and ecosystem modeling. In environmental sustainability, it helps in bioremediation studies, metagenomics, and climate-adaptive agriculture.

Skills and Competencies Required

A successful career in bioinformatics requires both **biological understanding** and **computational proficiency**.

Essential Skills

- Strong foundation in **molecular biology and genetics**
- Proficiency in **programming languages** (Python, R, Perl, C++)
- Familiarity with **bioinformatics tools and databases** (NCBI, UniProt, PDB, EMBL, Ensembl)
- Experience with **data analysis and visualization software** (Bioconductor, Cytoscape, PyMOL)
- Knowledge of **AI/ML techniques** for biological prediction models
- Understanding of **Linux/Unix environments** and **cloud computing**

Desirable Soft Skills

- Analytical thinking and problem-solving
- Teamwork and communication in multidisciplinary research
- Scientific writing and data interpretation

Career Opportunities in Bioinformatics

Academic and Research Careers

Researchers and academicians work in universities, national laboratories, and research centers. Positions include **Research Scientist**, **Bioinformatics Analyst**, and **Postdoctoral Fellow**, focusing on genomic analysis, structural modeling, or algorithm development.

Pharmaceutical and Healthcare Industry

Bioinformatics professionals contribute to **drug discovery**, **clinical genomics**, **precision medicine**, and **pharmacogenomics**. Companies such as **Pfizer, AstraZeneca, and Roche** actively employ bioinformaticians for computational drug design and biomarker validation.

Agricultural and Environmental Sectors

Bioinformatics experts are employed in **seed companies**, **biotech firms**, and **environmental agencies** for genome-based crop improvement, pest resistance research, and ecological data analysis.

Data Science and Artificial Intelligence

The rise of **AI in biosciences** has created new roles such as **Bioinformatics Data Scientist**, **AI Genomics Engineer**, and **Computational Biologist** in startups and tech companies integrating biology and machine learning.

Government and Public Sector

National agencies like **ICMR**, **DBT**, **CSIR**, **NIH**, and **NCBI** hire professionals for policy development, bioresource management, and public health genomics.

Global Job Trends

The demand for bioinformatics professionals has increased substantially due to the explosion of biological data. According to the U.S. Bureau of Labor Statistics (2024), bioinformatics-related roles are expected to grow by more than 15% over the next decade. In India, the Department of Biotechnology (DBT) has launched several initiatives to promote bioinformatics training and entrepreneurship through BTISNet and Skill India programs.

Major Recruiters Include

- Pharmaceutical companies (Novartis, Merck, Biocon)
- IT and AI-based firms (TCS, Infosys, NVIDIA)
- Research organizations (NCBI, EMBL-EBI, JNU, IISc, IITs)

Future Scope of Bioinformatics

Integration with Artificial Intelligence

AI and deep learning have revolutionized the field by enabling predictive models for **protein folding, drug response, and disease risk**. Tools like **AlphaFold** and **ChatGPT for Genomics** mark a new era of computational biology.

Personalized and Precision Medicine

Bioinformatics supports **genome-guided therapy** and patient-specific treatment strategies. Genetic profiling is increasingly used to design targeted drugs for cancer, diabetes, and neurological diseases.

Synthetic Biology and Biotechnology

Bioinformatics aids in **gene circuit design**, **metabolic engineering**, and **biofuel production**, contributing to sustainable industrial biotechnology.

Green and Environmental Applications

Bioinformatics tools are employed in **metagenomic studies**, **wastewater treatment**, and **bioindicator species analysis**, supporting global sustainability efforts.

Genomics and Proteomics

Bioinformatics is fundamental to understanding genetic information and protein functions. Applications include:

Genome assembly, annotation, and comparative genomics.

Identification of biomarkers for diseases.

Functional and structural analysis of proteins.

Drug Discovery and Development

In pharmaceutical industries, bioinformatics aids in:

Target identification and validation.

Drug design using computational docking and virtual screening.

Analysis of pharmacogenomics data for personalized medicine.

Agriculture and Food Security

Bioinformatics is used to improve crop yield and resilience by:

Identifying genes for stress tolerance and pest resistance.

Analyzing genomes of agricultural species for breeding programs.

Ensuring food safety through microbial genomics.

Personalized Medicine

With the advent of precision medicine, bioinformatics enables:

Tailoring treatments based on individual genetic profiles.

Understanding disease mechanisms through transcriptomics and epigenomics.

Predicting drug response and adverse effects.

Environmental and Evolutionary Studies

Bioinformatics contributes to studying biodiversity and ecosystem health by:

Analyzing metagenomic data of environmental samples.

Understanding evolutionary relationships through phylogenetic analysis.

Monitoring microbial communities in ecosystems.

Career Opportunities in Bioinformatics

Academia

Roles: Research Scientist, Professor, or Bioinformatics Instructor.

Responsibilities: Conduct research, publish findings, and teach bioinformatics courses.

Industry

Pharmaceutical and Biotech Companies: Roles include Bioinformatics Scientist, Data Analyst, and Computational Biologist.

IT and Big Data Companies: Positions such as Bioinformatics Software Developer and Data Scientist.

Government and Healthcare

Public Sector Research Organizations: Bioinformatics researchers contribute to national and global health initiatives.

Hospitals and Clinics: Roles in precision medicine and genetic counseling.

Entrepreneurship

Launching bioinformatics startups focused on software development, genomics services, or AI-driven biological analysis.

Required Skillsets

Technical Skills

Programming: Proficiency in Python, R, Java, or Perl.

Database Management: Knowledge of SQL and database systems like MySQL or MongoDB.

Bioinformatics Tools: Familiarity with BLAST, Clustal Omega, and tools for NGS data analysis.

Analytical and Statistical Skills

Proficiency in statistical methods and machine learning techniques.

Ability to interpret large datasets and derive meaningful insights.

Biological Knowledge

Understanding molecular biology, genetics, and biochemistry principles.

Awareness of biological databases like GenBank, UniProt, and KEGG.

Emerging Trends in Bioinformatics

Artificial Intelligence and Machine Learning

AI and ML are revolutionizing bioinformatics by enabling predictive modeling, drug discovery, and image analysis.

Cloud Computing

Cloud platforms facilitate handling large-scale genomic and proteomic data.

Single-Cell and Spatial Omics

Advances in single-cell sequencing are providing deeper insights into cellular heterogeneity and spatial organization.

Synthetic Biology

Bioinformatics is integral to designing synthetic genes and pathways for industrial and therapeutic applications.

Challenges and Future Directions

While bioinformatics offers vast opportunities, professionals face challenges such as:

Data Complexity: Handling and interpreting vast, heterogeneous datasets.

Interdisciplinary Knowledge: Bridging gaps between biology, computer science, and statistics.

Ethics and Data Security: Ensuring privacy and ethical use of genetic information.

Future directions include

Development of more efficient algorithms and tools for big data analysis.

Enhanced integration of multi-omics data for holistic biological insights.

Bioinformatics companies in India offering job opportunities for professionals in the field: For Example

Strand Life Sciences

Location: Bangalore

Focus: Genomics, precision medicine, and software development for healthcare.

Careers: Roles include bioinformatics scientists, data analysts, and software developers.

Website: strandls.com

MedGenome

Location: Bangalore

Focus: Genomics-based diagnostics, research, and personalized medicine.

Careers: Opportunities in NGS analysis, clinical bioinformatics, and research roles.

Website: medgenome.com

Genotypic Technology Pvt. Ltd.

Location: Bangalore

Focus: Genomics services, including sequencing and bioinformatics solutions.

Careers: Jobs in genomics analysis, molecular biology, and bioinformatics R&D.

Website: genotypic.co.in

SciGenom Labs

Location: Cochin

Focus: Next-generation sequencing, genomics research, and bioinformatics services.

Careers: Positions in genomics research, bioinformatics analysis, and software development.

Website: scigenom.com

Ocimum Biosolutions

Location: Hyderabad

Focus: Integrated bioinformatics solutions, software development, and genomics.

Careers: Roles in software development, data analysis, and bioinformatics R&D.

Website: ocimumbio.com

BioXcel Corporation

Location: Bangalore and US

Focus: AI-driven drug discovery and bioinformatics solutions.

Careers: Opportunities in computational biology, AI, and bioinformatics R&D.

Website: bioxcel.com

How to Prepare for a Career in Bioinformatics

Skills Required:

Proficiency in programming (Python, R, Perl).

Knowledge of bioinformatics tools and databases (BLAST, UniProt, etc.).

Experience in genomics, proteomics, or molecular modeling.

Qualifications:

A degree in bioinformatics, biotechnology, or computational biology, Microbiology, Biochemistry

Additional certifications or project experience in data analysis or machine learning.

ob Portals and Platforms:

Naukri.com

Linkedin

https://in.indeed.com/?from=gnav-jobsearch--indeedmobile

Conclusion

The career and scope of bioinformatics are expanding rapidly, driven by advancements in technology and the need for data-driven biological insights. With applications ranging from healthcare to agriculture, bioinformatics professionals have diverse opportunities to contribute to scientific innovation and address global challenges. Developing a strong foundation in computational and biological sciences is key to thriving in this dynamic field. Bioinformatics has transformed from a niche computational tool to a cornerstone of modern biological science. The integration of genomics, data science, and artificial intelligence has expanded its reach to nearly every scientific domain.For students and professionals, bioinformatics offers **diverse**

career pathways, strong **research prospects**, and global **industry demand**. As data generation accelerates, the ability to analyze and interpret biological information will remain one of the most valuable skills in the life sciences.

References

1. Mount, D. W. (2001). Bioinformatics: Sequence and Genome Analysis.
2. Karp, P. D., et al. (2018). Applications of Bioinformatics in Medicine and Biology. Journal of Computational Biology.
3. Sharma, A., et al. (2020). Bioinformatics Tools for Agricultural and Environmental Research. Frontiers in Bioinformatics.
4. Lesk, A. M. (2019). *Introduction to Bioinformatics* (٥th ed.). Oxford University Press.
5. Pevsner, J. (2015). *Bioinformatics and Functional Genomics* (٣rd ed.). Wiley-Blackwell.
6. Ranganathan, S., Gribskov, M., & Nakai, K. (2019). *Introduction to Bioinformatics* (٢nd ed.). Pearson Education.
7. Zhang, C., Zheng, W., Mortuza, S. M., Li, Y., & Zhang, Y. (2021). Deep-learning methods for protein structure prediction. *Frontiers in Molecular Biosciences, 8*, 771154.

List of Contributors

Dr Uma Kumari

Department of Bioinformatics (Bioinformatics Project and Research Institute, Noida -201301, UP

Dr Mohini Singh

Department of Life Sciences, School of Basic Sciences and Research, Sharda University, Knowledge Park III, India 201310

Devanshi Gupta

Department of Bioinformatics

Bioinformatics Project and Research Institute, Noida-201301, UP, India

Tanmay Bandbe

Department of Bioinformatics

Bioinformatics Project and Research Institute, Noida-201301, UP, India

Priya Arya

Department of Bioinformatics

Bioinformatics Project and Research Institute, Noida-201301, UP, India

Meenakshi Pradhan

Department of Bioinformatics

Bioinformatics Project and Research Institute, Noida-201301, UP, India

Vineeta Johri

Department of Bioinformatics

Bioinformatics Project and Research Institute, Noida-201301, UP, India

Glossary & Terminology

CHAPTER 1: Introduction of Bioinformatics

- Bioinformatics – An interdisciplinary field combining biology, computer science, mathematics, and statistics to analyze biological data.
- Computational Biology – Application of computational techniques to model biological systems.
- Genomics – Study of the complete DNA content (genome) of an organism.
- Proteomics – Large-scale study of proteins, including structure, function, and interactions.
- Transcriptomics – Study of RNA transcripts produced by the genome.
- Systems Biology – Integrated analysis of biological systems using computational models.
- Structural Bioinformatics – Study of three-dimensional structures of biomolecules.
- Personalized Medicine – Medical treatment tailored to individual genetic profiles.

CHAPTER 2: Sequence Format

- FASTA (Fast-All) – Text-based format for nucleotide or protein sequences.
- FASTQ (Fast-Quality) – Sequence format storing nucleotide sequences with quality scores.
- GenBank – Public nucleotide sequence database maintained by NCBI.
- Accession Number – Unique identifier assigned to a biological sequence.
- IUPAC (International Union of Pure and Applied Chemistry) – Organization defining nucleotide and amino acid codes.
- PAM (Point Accepted Mutation) – Substitution matrix for protein alignment.
- BLOSUM (Blocks Substitution Matrix) – Matrix used for scoring protein sequence alignments.
- VCF (Variant Call Format) – File format for storing genetic variation data.
- SAM (Sequence Alignment Map) – Text format for storing aligned sequences.
- BAM (Binary Alignment Map) – Binary version of SAM.

CHAPTER 3: Sequence Alignment

- Sequence Alignment – Arrangement of sequences to identify similarity.
- Pairwise Alignment – Alignment between two sequences.
- Multiple Sequence Alignment (MSA) – Alignment of three or more sequences.
- Global Alignment – End-to-end alignment of sequences.
- Local Alignment – Alignment of the most similar regions.
- BLAST (Basic Local Alignment Search Tool) – Tool for sequence similarity search.
- Needleman–Wunsch Algorithm – Global alignment algorithm.
- Smith–Waterman Algorithm – Local alignment algorithm.
- Dot Plot – Graphical method to compare sequences.
- Homology – Similarity due to shared evolutionary ancestry.

CHAPTER 4: Homology Modelling

- Homology Modelling – Prediction of protein structure based on known templates.
- Template – Known protein structure used for modelling.
- Swiss-Model – Web server for automated homology modelling.
- GalaxyWeb – Platform for protein structure prediction and refinement.
- Robetta – Server for protein structure prediction.
- RMSD (Root Mean Square Deviation) – Measure of structural similarity.
- Ramachandran Plot – Visualization of protein backbone angles.
- Protein Validation – Assessment of model quality and reliability.

CHAPTER 5: Computational Biology

- Mathematical Modelling – Representation of biological processes using equations.
- Simulation – Computational imitation of biological behavior.
- Network Biology – Study of biological interaction networks.
- Pathway Reconstruction – Mapping biological pathways computationally.
- Multi-Omics – Integration of genomics, proteomics, transcriptomics data.
- Data Integration – Combining heterogeneous biological datasets.

CHAPTER 6: Biological Databases

- NCBI (National Center for Biotechnology Information) – Central bioinformatics resource.
- EBI (European Bioinformatics Institute) – European bioinformatics data center.

- DDBJ (DNA Data Bank of Japan) – Japanese nucleotide database.
- UniProt (Universal Protein Resource) – Protein sequence and annotation database.
- PDB (Protein Data Bank) – Repository of 3D biomolecular structures.
- GO (Gene Ontology) – Standardized gene function vocabulary.
- KEGG (Kyoto Encyclopedia of Genes and Genomes) – Pathway database.
- PRIDE (PRoteomicsIDEntifications Database) – Proteomics data repository.

CHAPTER 7: Genomics and Proteomics

- Structural Genomics – Study of genome-encoded protein structures.
- Functional Genomics – Study of gene functions and interactions.
- Comparative Genomics – Comparison of genomes across species.
- Expression Proteomics – Study of protein expression levels.
- Protein–Protein Interaction (PPI) – Physical interaction between proteins.
- Proteome – Entire set of proteins expressed by a genome.

CHAPTER 8: Computer Aided Drug Designing (CADD)

- CADD (Computer Aided Drug Designing) – Computational drug discovery approach.
- Target Identification – Selection of biological targets for drugs.
- Molecular Docking – Prediction of ligand–protein binding.
- Virtual Screening – Computational screening of compound libraries.
- ADMET (Absorption, Distribution, Metabolism, Excretion, Toxicity) – Drug safety parameters.
- Pharmacophore – Structural features responsible for drug activity.
- Molecular Dynamics (MD) – Simulation of molecular movements.

CHAPTER 9: Artificial Intelligence & Machine Learning

- AI (Artificial Intelligence) – Machines performing human-like intelligence.
- ML (Machine Learning) – Algorithms learning patterns from data.
- Supervised Learning – Learning using labeled data.
- Unsupervised Learning – Learning without labeled outputs.
- Deep Learning – ML using neural networks with multiple layers.
- Feature Engineering – Selection of input variables for ML models.
- Model Validation – Assessment of predictive performance.

CHAPTER 10: Next-Generation Sequencing (NGS)

- NGS (Next-Generation Sequencing) – High-throughput DNA sequencing.
- Illumina – Sequencing-by-synthesis platform.
- PacBio (Pacific Biosciences) – Long-read sequencing technology.
- Oxford Nanopore – Real-time nanopore sequencing.
- Read Mapping – Aligning reads to reference genome.
- Variant Calling – Detection of genetic variations.
- RNA-Seq (RNA Sequencing) – Transcriptome sequencing method.
- Quality Control (QC) – Assessment of sequencing data quality.

CHAPTER 11: Bioinformatics in Food Biotechnology

- Food Bioinformatics – Application of bioinformatics in food science.
- Nutrigenomics – Study of nutrition–gene interactions.
- Metagenomics – Analysis of genetic material from microbial communities.
- Probiotics – Beneficial microorganisms for health.
- Functional Foods – Foods providing health benefits beyond nutrition.

CHAPTER 12: NGS with Biopython

- Biopython – Python library for bioinformatics analysis.
- Sequence Parsing – Reading biological sequence files programmatically.
- NGS Pipeline – Automated sequence analysis workflow.
- Python – High-level programming language used in bioinformatics.

CHAPTER 13: Structural Bioinformatics

- Secondary Structure – Alpha helices and beta sheets.
- Tertiary Structure – Three-dimensional protein folding.
- Quaternary Structure – Assembly of multiple protein subunits.
- Protein Folding – Process by which protein attains functional shape.
- Structure Visualization – Graphical display of biomolecules.

CHAPTER 14: Bioinformatics in COVID-19

- SARS-CoV-2 – Severe Acute Respiratory Syndrome Coronavirus-2.
- Viral Genomics – Study of viral genetic material.

- Vaccine Design – Computational development of vaccines.
- Drug Repurposing – Using existing drugs for new diseases.
- Network Pharmacology – Drug–target interaction network analysis.

CHAPTER 15: Career and Scope of Bioinformatics

- Bioinformatician – Scientist specializing in biological data analysis.
- Research Scientist – Professional conducting scientific research.
- Industry Bioinformatics – Application in pharma, biotech, healthcare.
- Academic Bioinformatics – Teaching and research in universities.
- Skill Sets – Programming, statistics, biology, data analysis.

Appendix

Appendix A: List of Common Bioinformatics Databases

- This appendix provides a consolidated list of widely used biological databases referenced throughout the book.
- NCBI (National Center for Biotechnology Information) – Central repository for nucleotide, protein, genome, and literature databases.
- GenBank – Public nucleotide sequence database.
- UniProt (Universal Protein Resource) – Protein sequence and functional annotation database.
- PDB (Protein Data Bank) – Repository of experimentally determined 3D biomolecular structures.
- KEGG (Kyoto Encyclopedia of Genes and Genomes) – Pathway and systems biology database.
- Gene Ontology (GO) – Controlled vocabulary for gene and protein functions.
- ENSEMBL – Genome browser providing integrated genomic data.
- PRIDE (PRoteomicsIDEntifications Database) – Proteomics mass spectrometry data repository.
- GEO (Gene Expression Omnibus) – Functional genomics and gene expression database.

Appendix B: Common Bioinformatics File Formats

- This appendix summarizes major file formats used in bioinformatics analysis.
- FASTA – Sequence format for nucleotide and protein sequences.
- FASTQ – Sequence format including quality scores.
- GenBank (.gb/.gbk) – Annotated nucleotide sequence format.
- PDB – Protein structural coordinate format.
- GFF (General Feature Format) – Genomic annotation format.
- VCF (Variant Call Format) – Format for genetic variant data.
- SAM/BAM – Alignment formats for sequencing reads.

Appendix C: Frequently Used Bioinformatics Tools

- A brief list of computational tools discussed in the book.
- BLAST – Sequence similarity search tool.
- Clustal Omega – Multiple sequence alignment tool.
- Swiss-Model – Homology modelling server.
- GalaxyWeb – Protein structure prediction and refinement platform.
- Robetta – Protein structure modelling server.
- Biopython – Python library for bioinformatics analysis.
- STRING – Protein–protein interaction database and analysis tool.

Appendix D: Mathematical and Statistical Terms Used

- RMSD (Root Mean Square Deviation) – Measure of structural similarity.
- Scoring Matrix – Matrix used to score sequence alignments.
- Gap Penalty – Penalty applied for insertions or deletions in alignments.
- Feature Engineering – Selection of relevant variables for machine learning models.
- Model Validation – Evaluation of predictive performance of computational models.

Appendix E: Ethical and Regulatory Considerations

- This appendix highlights ethical aspects related to bioinformatics research.
- Data privacy and genetic data protection
- Ethical use of AI and machine learning
- Responsible sharing of genomic data
- Compliance with institutional and international guidelines
- Reproducibility and transparency in computational research

Appendix F: Suggested Learning Pathway

- For students and early-career researchers:
- Fundamentals of Bioinformatics and Sequence Analysis
- Biological Databases and Data Retrieval
- Sequence Alignment and Homology Modelling
- Genomics, Proteomics, and NGS
- Structural Bioinformatics and Drug Design
- AI and Machine Learning in Biology
- Applied Case Studies and Research Projects

Appendix G: Software and Programming Skills

- Recommended technical skills for bioinformatics professionals:
- Programming Languages: Python, R
- Operating Systems: Linux/Unix
- Data Handling: CSV, FASTA, FASTQ, VCF
- Visualization: PyMOL, Chimera
- Version Control: Git
- Workflow Tools: Galaxy, command-line pipelines

Appendix H: Abbreviations and Acronyms (Selected)

- AI – Artificial Intelligence
- BLAST – Basic Local Alignment Search Tool
- CADD – Computer Aided Drug Designing
- NGS – Next-Generation Sequencing
- PPI – Protein–Protein Interaction
- RMSD – Root Mean Square Deviation

Index

E

F

G

H

I

K

L

M

N

P

R

S

T

V

D